How to Do *Everything* with

Digital Video

DIRECTOR

PRODUCTION

OUR FAMILY MOVIE

DATE	SCENE	TAKE
9-5	12	9

Frederic Jones

McGraw-Hill/Osborne

New York   Chicago   San Francisco   Lisbon
London   Madrid   Mexico City   Milan   New Delhi
San Juan   Seoul   Singapore   Sydney   Toronto

McGraw-Hill/Osborne
2600 Tenth Street
Berkeley, California 94710
U.S.A.

To arrange bulk purchase discounts for sales promotions, premiums, or fund-raisers, please contact **McGraw-Hill**/Osborne at the above address. For information on translations or book distributors outside the U.S.A., please see the International Contact Information page immediately following the index of this book.

How to Do Everything with Digital Video

1234567890 CUS CUS 01987654321

ISBN 0-07-219463-4

Publisher	**Project Manager**	**Computer Designer**
Brandon A. Nordin	Deidre Dolce	Kate Kaminski, Happenstance Type-O-Rama
Vice President & Associate Publisher	**Freelance Project Manager**	
Scott Rogers	Laurie Stewart	**Illustrator**
	Copy Editor	Jeffrey Wilson
Editorial Director	Lunaea Weatherstone	**Series Design**
Roger Stewart		Mickey Galicia
	Proofreader	
Acquisitions Editor	Sarah Kaminker	
Megg Bonar		
	Indexer	
Acquisitions Coordinator	Jack Lewis	
Tana Diminyatz		

This book was composed with QuarkXPress 4.11 on a Macintosh G4.

CANON and XL1 are registered trademarks of Canon Inc. All rights reserved. Used by permission.

To my father, Robert F. Jones,
who has always believed I could do anything.

About the Author

Frederic Jones attended Ph.D. programs at Florida State University and International College. He has served as CEO of Eclat, Inc., an electronic publishing company specializing in multimedia electronic product information systems, and as CTO/General Manager of 911 Entertainment, Inc., an Internet music and entertainment company. At Eclat, he was responsible for building both the business strategy and the technology to deliver extensive intelligent catalog systems for GE, Westinghouse, Herman Miller Corp., Steelcase, Moen, Westinghouse Corp., Lightolier Corp., Sigma Design, Inc., and others. At Ebook and Jones & Jones Multimedia, where he was CEO, he was responsible for creating a national/international CD-ROM publishing business focusing on consumer products and games. These included over 30 products with millions of units sold in Europe, North America, and Asia.

At 911 Entertainment, Dr. Jones, who was the General Manager and Chief Technology Officer, was responsible for design and development of a sophisticated Internet e-commerce and webcasting site to support extensive music CD publishing and marketing. He has consulted on Internet marketing and e-business with such companies as Microsoft, Cybercash, Virgin, GTE, Sony Entertainment, FOX Entertainment, MCA Records, eMusic, Nanospace, SonicSolutions, and others. He has consulted on distance learning technology and course development with City & Guilds of London, Tallahassee Community College, Florida Community College Distance Learning Consortium, Pacific School of Religion, and DigitalThink, Inc.

Dr. Jones has written hundreds of periodical articles and 23 books, many of which are in the area of electronic information and architecture.

Dr. Jones is an accomplished videographer, 3D artist, and animator and has developed and animated major game titles (such as *Beyond Time*, created by Jones & Jones Multimedia, Inc., and published by Virgin/Panasonic/DreamCatcher) using Premiere, After Effects, 3D Studio, Lightwave, TrueSpace, and other products. His architectural and lighting design background brings a rich resource to his video and animation projects.

He would be glad to hear from you at fjones@acousticsquare.com.

Contents

viii   Contents

Acknowledgments

There are many people who have helped in the writing and production of this book who need special thanks. The first and most important is my wife and partner, Judith K. Jones, who edited and agonized over the chapters with me. She also wrote most of Chapters 18 and 19. She is a very experienced video and multimedia producer, artist, and author and without a doubt, could have written the entire book better than I. I would also like to credit my daughter Mary Atchison for her help in working with the digital cameras, shooting sample files, and many other invaluable tasks.

Thanks also to the kind people at SCM Microsystem, particularly Kimberly Blackledge and Dirk Peters, who provided me with software and hardware products to review and write about. Their product information is available at http://www.dazzle.com/. Greg Kochaniak, who provided the Hypercam software and other materials, was a great help. Their product information is available at http://www.hyperionics.com/. Thanks also to Erin Roche of Weber Shandwick Worldwide, the PR firm that provides assistance to authors writing on Microsoft multimedia products, for wrestling an advance copy of Windows XP for me to use. Dave Chaimson at Sonic Foundry helped with copies of Sound Forge, ACID, and more as well as technical feedback. Their product site is http://www.sonicfoundry.com/.The support department at RealNetworks provided a variety of their product for review and inclusion. Their website is http://www.realvideo.com/. Last, but not least, thanks to my editor, Megg Bonar, who showed amazing restraint and patience with me as I learned the McGraw-Hill/Osborne process. I am sure that I have forgotten someone, but they are all very much appreciated.

Introduction

Video is about one thing: telling a story. Whether it is a story of events, people, or ideas and relationships, video is a fascinating and facile media for the expression of story. With the ease and versatile technology of digital video, you can now enter this special world of storytelling and, more important, excel at it.

Welcome to the wonderful world of digital video. You have just purchased a digital video camera or are thinking of doing so, and you want to know about using it creatively and well. You have come to the right place. This is a book for people who want to do more with their cameras than just point, shoot, and toss the videotapes in a box in the closet. We all know people who have stacks of hastily labeled videotapes—or even worse, anonymous tapes containing no-one-remembers-what, and usually with good reason. Many people associate other people's self-shot videotapes as a major source of boredom. Endless sequences of wobbly tape, poorly lighted, excruciating sound, shapeless content—these are the hallmarks of amateur video, right? Wrong! I'll tell you how to avoid all the grisly mistakes of amateur video, and how to shoot effective and compelling video. In this book, we are going to create great video projects that you will be proud of and your friends and family will enjoy watching.

There are more possibilities for using digital video than I can present in this book. You can document family history, immortalize special milestone events and anniversaries, create promotional videos for organizations, make the perfect wedding video, tell stories, record vacation videos that will remind friends of National Geographic, and, of course, tape Bob Jr.'s first birthday party. What's more, you can create these projects and view them on tape, from CD-ROM, or over the Internet. Dreaming of telling stories on film, of producing a mini-movie? Read on, I'll tell you how.

Whatever the end product is destined to be, the most essential thing is to shoot your video with that end in mind. You must be well organized, well prepared, and accustomed to thinking in terms of the elements of video. Great projects don't just happen, they are designed and planned with an eye to detail. You will learn how to plan your projects so you are certain to get the shots you need to tell the story. Once you develop an eye and an ear for video possibilities, you will learn how to polish your work with editing. Editing is not about technical effects, it is about the communication of ideas and emotions. It is about taking raw footage and assembling, trimming, and enhancing it so it accomplishes its goal—telling the story! The addition of music, special effects, titles, narration, and a host of other finishing touches will give your projects exactly that—finish. Stop thinking of videotape as start-stop film clips of interest to no one, not even you! Start thinking of the endless possibilities for excellent and fascinating digital video production.

Organization of the Book

In **Part I**, I begin with the basics of camera use and features. I talk about how to hold the camera so you achieve steady video, how to pan and zoom, when to use both, and when not to use either. I discuss the basic types of shots and how to achieve them. When you're ready to shoot, I explain how to plan for and capture the elements you need to shoot in order to tell your complete story or make your points in a finished video. I also cover how to ensure basic video and sound quality.

In **Part II**, I cover the basics of assembling the scenes and elements captured into a finished and edited video. Remember, it is almost impossible to extract a comprehensible story from randomly shot videotape, as you almost always are lacking some important elements or important elements were shot without emphasizing the key features needed for good storytelling. It is impossible to fix many filming mistakes at the time of editing (such as removing the airplane noise that drowns out little Bobby's Happy Birthday song or recovering Aunt Clara's facial expressions if she was filmed with her back to the sun).

In **Part III**, I cover ways your video can be shared with others, by tape, CD-ROM or the Internet.

In **Part IV**, I lead you through the process of planning, shooting, editing, and distributing a complete mini-project. This will give you a "cookbook recipe" for specific popular projects as well as provide you with a model for the practical application of the techniques introduced in Parts I through III.

In **Part V**, the appendix introduces the video and other multimedia features of Microsoft Windows XP. It includes a brief tutorial on the transfer of digital video from digital camcorder to computers and the use of Windows XP tools for editing and distributing digital video via CD-ROM, email, and the Internet.

Part I

Digital Video Cameras and Camcorders

Chapter 1

Learn the Basics of Your Digital Video Camera

How to...

- Understand the uses and potential of digital video camcorders
- Select the best digital camcorder for your needs
- Understand the basic features of your camcorder
- Understand the differences between American and European video standards
- Operate the basic features of your camcorder
- Connect your camcorder to external equipment and computers

We live in a visual world. More than any other sense, vision dominates our culture. Visual media of all kind bombards us, from still images on billboards and in publications to television and movies. Nothing gives a person with a desire for creativity any greater sense of power than a video camera and the ability to "make movies." Digital video camcorders, with their capacity for editing on the home computer, enable us to move from the old and relatively crude level of home movies to finished and well-made visual records of events, personal and family history, business projects, and, of course, entertainment and art films. With a digital video camera, we can all be producers of exciting and fun video projects. Fortunately, as technology has expanded the availability and technological capability of digital video cameras and camcorders, their ease of use has also increased. Learning the basics will get you quite a way down the road toward the creation of far more creative and sophisticated video projects than you ever imagined.

Digital Video Cameras and Digital Video Camcorders

It is common, though misleading, to refer to a video camcorder both as a camcorder and video camera, and this will continue after this book, but let's take a moment to point out the difference between them. Basically, a *video camera* consists of:

- A lens
- A semiconductor device or image chip that converts the images concentrated by the lens to electrical signals
- A series of electronic circuits that process the electrical signals from the image chip into television signals that can be viewed on a TV screen or recorded by a video recorder or VCR

A *camcorder* is a video camera with a video recorder built into it. A *digital video camera* is a video camera that has additional electronic circuits added to convert the basic television signals (consisting of analog signals) into digital signals. A digital video camcorder also has a video recorder capable of recording digital rather than standard analog television signals.

From the beginning of television, both cameras and camcorders were analog devices. In the 1970s, digital audio and video were invented and began to find their way into television, first as a means of adding special effects to programs, then as an editing device, and finally as cameras, camcorders, monitors, and view screens. Analog camcorders include VHS, SVHS, VHSC, 8mm, and Hi8 in consumer formats and many other professional formats. In the digital world, consumer formats include DV and Digital-8. I will explain these a bit more in the sections that follow.

NOTE *Analog signals are continuous electrical signals that vary in voltage and frequency. These signals are recorded to tape or disks as exact representations of those continuous signals. Digital technology records analog signals as a series of momentary "samples" of the voltage and frequency represented as a list of numbers. The human ear and eye perceive images and sound as continuous analog signals, so the digital signals that record audio and video must be converted back to analog to be displayed on a television or computer monitor for human viewing.*

The camcorder is the typical video device that consumers and most professionals use to film video. You will probably encounter video cameras as attachments to your computer for capturing web video. I'll touch on these briefly in a later chapter, but this book will concentrate on the digital camcorder. Figure 1-1 shows a typical miniDV camcorder.

FIGURE 1-1   A typical miniDV camcorder

The DV Format

The most common consumer digital video format is the *DV* (*digital video*) format. This is a standard that has been adopted worldwide for consumer camcorders, recorders/VCRs, and other video devices. It uses a small cassette tape to record the signals and includes a certain standard for recording both audio and video signals so they can be properly exchanged between equipment from different manufacturers.

The Sony Digital-8 Format

Sony has developed a variation on the DV format that uses the same electronic and signal standards, but records the signal on a standard 8mm or Hi8 analog videocassette. The Digital-8 equipment also plays legacy analog 8mm and Hi8 tapes and can translate them into digital form when copying to other digital devices. Although this format is unique to Sony, the signal can be copied to and from DV equipment and can be interchanged in the process of editing on the computer. Sony also manufactures DV camcorders and equipment as well as a wide range of professional formats, both analog and digital.

International Video Standards

In the United States (and some other countries), a standard analog video signal consists of 525 lines and 30 frames per second. This standard is called the NTSC standard, for the National Television Standards Committee, which created it. In most of Europe, the standard is called PAL; this standard has 625 scanned lines and 25 frames per second. In France and a few other countries, there is a third standard called SECAM, which is similar to PAL. These standards are incompatible with one another. In other words, videotapes or broadcast signals in one format can't be viewed on TVs or VCRs that utilize the other formats. In addition, there are new high-definition analog and digital television standards that will be supplanting the existing standards over the next few years. For now, standard television and both DV and analog consumer camcorders use NTSC, PAL, or SECAM standards. In the United States, it will be NTSC exclusively. Unless you are exchanging tapes with friends overseas, you don't need to be concerned about these differences. If you do exchange tapes overseas, many video stores in large cities provide format conversion services. Most digital video editing systems can edit and output NTSC, PAL, and SECAM as well as many other digital formats if properly set. Analog monitors, recorders, and cameras cannot be switched.

Time Code

Time code is a function of camcorders that specifically identifies video frames and audio signals by time. In the recorded time code, each frame of visual and audio material has a unique identifier, or address. A time code consists of four two-digit numbers, which represent hours, minutes, seconds, and frames on a 24-hour clock. In this way, frame-accurate, repeatable edits are possible. You can set and view time codes in the viewfinder of most camcorders and/or on the LCD readout on the side of the camcorder. These can be set and noted when you are shooting so that you can accurately rewind and return to shots or refer to them when editing. The time code is recorded on the DV tape and goes with the tape from camcorder to camcorder. Normal time code is called *non-drop-frame time code* and records the exact length of the video sequence. (Drop-frame time code is explained below.) Time code functionality is only available on high-end consumer (commonly called *prosumer*) and professional camcorders; most consumer models do not have true time code.

Drop-Frame Time Code

While the novice user may not need to use drop-frame time code, it is something more advanced users may want to consider. Generally, NTSC time code is considered to run at 30 frames per second. In reality, for a number of technical reasons, it actually runs at 29.97 frames per second. *Drop-frame time code* runs accurately at the 29.97 frames per second. This is accomplished by skipping or dropping frame numbers 00 and 01 at the beginning of each minute, except at the ten-minute mark. No actual video frames are dropped—only the respective frame numbers are omitted from the time code sequence. It is important for many editing purposes that you set the edit software to specify drop-frame time code. This is not a setting that is used on camcorder setups for consumer camcorders.

Technical Specifications for Consumer DV and Digital-8

There are a variety of technical specifications in use and many technical terms you're likely to encounter in your search for and use of a digital video camera. Rest assured, it's not necessary for you to understand each of the technical bits. I've included a list here of some terms you'll see, and these will be explained as we go, to the extent you need to know about them to use your digital camcorder effectively.

Video recording   NTSC, PAL, or SECAM formats

Audio recording   PCM digital sound, 16-bit (48 KHz with two channels), or 12-bit (32 KHz with two channels or 32 KHz with four channels simultaneously)

Tape formats   miniDV or Digital-8

Tape speeds

   SP   .75 inches per second (18.81 mm per second)

   LP   .5 inches per second (12.56 mm per second)

Maximum recording time

> **SP**   60 minutes with a 60-minute cassette
>
> **LP**   90 minutes with a 60-minute cassette

DV outputs   IEEE 1394, also called FireWire or I-Link (by Sony)

What to Look for When You're Buying a Digital Video Camera

Now that you know what the definitions and specifications of a digital video camcorder are, you will put down the book and run out to buy one, if you haven't already. Wait just a moment! There are a few things you might want to consider.

The quality of digital video or miniDV (the standard consumer digital format) camcorders is, at this point, very high. They are manufactured by a number of reputable companies, including Sony, Panasonic, JVC, Canon, and others. The three brands with the widest range of low-end consumer to professional models available in this format are Sony, Canon, and Panasonic. These companies are the leaders in professional video gear as well.

The best way to shop is to consider how you intend to use your camcorder. If you are going to use it only to record your vacation and the kids' birthdays, you might consider one of the inexpensive models. If you are making amateur movies or shooting weddings on the weekends, a mid-price to high-end model will be best.

Here are a few things to consider:

Still photos   Many models come with built-in digital still cameras. Multiple-use devices always scare me. Neither feature ever seems as good as on the dedicated models. If you want to select a dual-use model, the ones that record still images onto memory sticks rather than to DV tape seem to give the best results and don't add wear and tear to the camcorder transport.

Color depth   If you are doing professional work, you should consider a model with three CCD (charged coupled device) chips in the camera. This gives much greater color depth to the recorded images. This quality is essential if you are transferring video to professional settings. These cameras, of course, cost much more. The single CCD cameras are still virtually identical in image quality for typical home use.

External microphone   The best camcorders allow for use of external microphones. This will allow you to avoid the slight whir caused by the internal microphone recording the transport noise. It will also allow for more accurate microphone placement when recording interviews and such. Distant microphone placement picks up more background noise and can mask spoken words.

Zoom   The quality of optical zoom is much better than digital zoom. Most camcorders offer both, but make sure that you have at least a basic optical zoom. Go for a larger zoom ratio (20x is better than 10x). If the camcorder you choose has both types of zoom, that's even better.

Image stabilization   If you are going to do a lot of handheld shooting, consider a camcorder with image stabilization. This is a built-in circuit that records a fraction of a second of image ahead and compensates for camera jiggle. It is a remarkable feature. The smaller camcorders are almost useless without image stabilization or a tripod.

Batteries and accessories   Compare battery life *and* battery cost when you are choosing a camcorder. An inexpensive camcorder can turn expensive fast when you go to buy extra batteries. The same consideration should be paid to accessories you might plan to buy. Check out both availability and cost of accessories, for example, extension lenses, remote control devices, etc., before you purchase the basic camcorder.

High-tech gadgets   Low-light and infrared recording are neat features to experiment with, but they generally are only a novelty except for spies.

Using analog tapes   If you have a Hi8 or 8mm analog camcorder and/or have a stash of legacy tapes, you might consider a Sony Digital-8 camcorder. The Digital-8 format can allow you to utilize those older tapes, transfer them to your computer, and mix shots with your Hi8 analog camcorder when you need a two-camera shot. The digital signals are completely compatible with standard miniDV and can be directly transferred to miniDV cassettes later with no loss of quality.

There are many other things you might consider in making your choice. Good sources of additional information are some of the excellent amateur and professional video magazines and the myriad of video-themed websites on the Internet.

Basic Operation of Your Camcorder

The main things you will want to do with your digital camcorder are tape, record, and play back video images and sound. Fundamentally, you do this by removing the lens cap, turning on the power, inserting a tape, pointing the camcorder at the subject, and pressing the Record button. I'll explain how to do these in more detail in the sections that follow.

Camcorder Features

This section will cover the key features commonly found on digital video camcorders. Specific features may or may not be found on a particular brand or model, and the control and specification will vary. When I was doing research for this book, I examined more than 25 camcorders and discovered that operation of some features, even very basic ones, is often not very "user friendly," and it was necessary to refer to user manuals. In light of this, please read the instruction booklet that accompanied your camcorder for specific details and operating instructions.

The Lens

The primary window of the video camcorder is the lens. A professional camcorder comes standard without a lens (which can cost as much as the camcorder itself). Most consumer camcorders come with a built-in lens that cannot be changed. One exception is the Canon XL1, which comes with a standard lens and can be fit with optional lenses. This camcorder can also be used with standard Canon 35mm film camera lenses with a special adapter. This and similar camcorders are designed for the prosumer markets and are frequently used by professionals because of their smaller size and light weight. In the basic consumer camcorder, you get what they provide, and the quality is generally good.

The lens directs light onto a *charged coupled device* (CCD), which is a semi-conductor chip that converts the light into electrical impulses that can be processed by the camcorder's circuitry. Low-end camcorders come with one CCD chip that processes all three primary colors. More expensive camcorders are equipped with three CCD chips and process the primary colors independently. The three-chip system results in higher quality images, which can be used more freely in professional and TV broadcast situations. While it is good to have the highest quality possible, the practical difference for amateur video is insignificant. Your video quality would be better served by buying a more expensive tripod than shelling out the extra money for three-chip camcorders.

In addition to color quality, CCD chips also offer a range of light sensitivity ratings, generally expressed in lux numbers. A *lux* is a metric measure of light levels. A higher number denotes a brighter light. Camcorders are best with the lowest possible lux number, which indicates that it will record images in situations with low light levels. This doesn't mean you can capture the highest quality video images in very low light, as the nature of low light creates an image that is most likely diffused and has indistinct shadows and modeling. Low-light shooting will record information and adequate video images. You still need to pay attention to good, adequate lighting when you are shooting high quality scenes.

Zoom lenses are standard on digital camcorders. Zoom lenses allow the operator to change the focal length of the lens from wide to narrow to record a shot with more in the picture (a wide shot) or a close-up on a particular subject (a narrow or tight shot). These lenses come in both optical and electronic zoom. Many camcorders are equipped with both. The optical zoom achieves the best quality result and should be used for high quality shots.

> TIP
>
> *Use zoom sparingly as an effect. Be sure to practice before using it on important shots. Zoom at a constant speed and not too fast.*

Setting the F-stop *F-stop* is a measure of how much light is being let through the lens. As in the human eye, the actual aperture that controls this is called the *iris*. Most lenses have marked settings ranging from *f*1.8 to *f*16. For example, an iris setting of *f*8 means that the lens is open to 1/8th of its total diameter. The smaller the f-stop number, the more light is passed through to the lens.

A rough technique for determining when there is enough light is to set the camera to auto-exposure or auto-iris. In this mode, the camera adjusts the iris automatically. Auto exposure tracks only the brightest part of the scene, so zoom out wide and pan around a bit (move the camera to take in more of the scene) to find an average reading. The iris setting should read about *f*5.6 or higher. An f-stop reading lower than *f*3.2 means the lens is almost totally open, taking in as much light as it can. This may result in a dark picture, loss of fine detail, colors not having rich saturation, and a reduced depth of field.

Setting the Depth of Field   Depth of field is the range of distance, closer to or further away from the camera lens, within which objects in a scene are in focus. One of the factors in determining depth of field is the iris setting. Low f-stop (for example, *f*2) numbers produce a short depth of field, and larger f-stop (such as *f*16) numbers provide a greater depth of field. Another factor is focal length. When a lens is zoomed to a wider field of view, objects within the scene will be in focus over a greater depth of field. Similarly, when it is necessary to zoom in on a scene (narrow the field of view), especially in low light where the iris must be set to a low f-stop value, it will be more difficult to achieve good focus.

Should You Use Filters?

Lenses are normally built to hold filters, pieces of glass that have been tinted, treated, or designed to hold some material between the subject and the camera. The effects may be subtle or stunning, and are more "organic" than post-production effects added later. Effects include polarization, diffusion, fog, star pattern, and gradated colors. At the very least, always use a skylight or UV filter. These are important for two reasons. First, they cut down on ultraviolet light entering the camera, creating a bluish tinge in shadows. They also act as inexpensive protection for that very expensive lens, keeping dust, dirt, and fingerprints off the main glass.

 *When cleaning a lens, never blow on it. That just gets more dirt stuck down past the seals. Always use proper lens cleaning fluid and paper. And never, ever use alcohol, which can remove the special coatings on many lenses.*

White Balance

A variety of features and controls can add greatly to the cost of a camera, but ultimately make a significant difference in convenience of operation and quality of results. One example is *white balance*. This is the way you tell the camera what type of light it is shooting in and, therefore, how it should render colors.

The human eye and brain are able to adjust and compensate in a variety of lighting conditions. Cameras just see what is there, in an objective electronic fashion. Different light sources, and lamps using different types of elements, produce different colors of light. Tungsten light is yellow, while fluorescent light is blue. Daylight is blue during high noon on a cold November day, but orange at sunset in August. We refer to such different lights as being "cool" or "warm."

The ability to control the white balance produces colors that seem natural and avoids the bluish look of much home video. Even consumer camcorders come with simple lighting controls, often just "indoor" or "outdoor." High-end cameras allow you to set the white balance for a particular lighting setup. When you focus on a piece of white paper, the camcorder reads the degree of "warm" or "cool" and adjusts accordingly.

If you have a manual white balance control, it is also possible to fool the camcorder into accepting a different color scale for a special effect. For instance, by white balancing to a "cool" light source and then using tungsten lighting during the shoot, the video will be "warm"— that is, rich in yellows and oranges. These settings can also be stored for future use in similar circumstances.

The Eyepiece and/or Video Monitor

All camcorders are equipped with an eyepiece and/or a built-in video monitor screen, as shown in Figure 1-2. The eyepiece consists of a very small video screen, usually black and white, with a magnifying lens in front of it. This is designed to work like a standard film camera viewfinder for framing your shots and showing setup information as overlays on the screen (for example, "Record ON" and the time code). Many recent camcorders also have small built-in flat-screen color monitors for viewing playback or more easily viewing the scene being taped when the camera is mounted on a tripod. A third possibility is to use an external monitor plugged into the video output of your camcorder. Small monitors are available as accessories and attach to the hot-shoe clip on the top of the camcorder. Small portable TV sets with AV inputs are great for this purpose. I have a five-inch color Magnavox AC/battery-powered TV that serves this purpose well. It is also better for playback from the camcorder in the field because it has a bigger screen and a built-in speaker for the sound.

The Microphone

Virtually all camcorders come with a built-in stereo microphone, as shown in Figure 1-3. These are adequate for casual shooting, particularly when the camcorder is close to the sound being recorded (five or six feet maximum unless it is a concert or sporting event where the roar of the crowd is what you are after). For other applications, you will want to use an external microphone plugged into the external microphone input of your camcorder. If you are narrating while shooting, a head clip microphone or a microphone/headphone combination is excellent. For interviews, wireless or wired clip-on microphones are a good choice. Tabletop microphones or a cable from a public address mixer board is good when recording speeches or lectures. We will cover microphone selection and placement in Chapter 5.

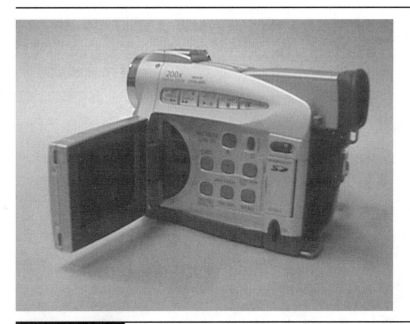

FIGURE 1-2   Close-up of on-camera monitor

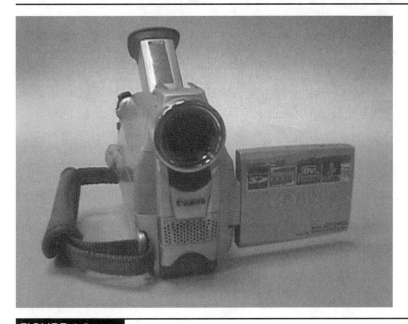

FIGURE 1-3   Typical microphone

> **TIP** *Most camcorders have no manual audio level controls, only automatic level settings. These will do fine. If you do have manual audio level controls, you will want to monitor the sound with a pair of headphones plugged into the headphone output jack while shooting to make sure the level is properly set.*

The Video Recorder Transport

The *video recorder transport* (Figure 1-4) is the VCR record and playback mechanism and the control buttons that operate it. It functions in exactly the same way that you are used to on your home VCR or other similar device.

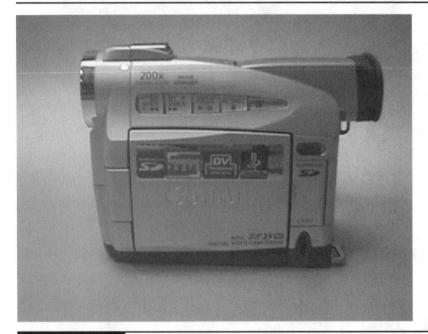

FIGURE 1-4   Close-up of transport cassette holder

The controls usually include Fast Forward, Forward, Play, Pause, Rewind, and Fast Rewind. Sometimes a red Record button is included. On most camcorders, this is a separate button on top as well as being associated with the Start and Stop buttons on the side grip of the camcorder.

Power Supply

Most camcorders come with both battery and external power sources. The external power sources often include direct connections to the camcorder to enable you to operate it without a battery and to charge the battery. Some camcorders come with both a power supply (see Figure 1-5) and a separate battery charger. Most camcorders also have optional fast chargers or multiple battery charging devices. Most camcorders can be optionally powered by car battery cables that plug into an automobile cigarette lighter or power outlet.

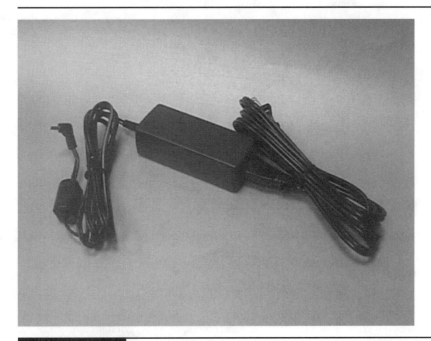

FIGURE 1-5   A typical power supply

Signal Input and Output Connectors

Camcorders come with a variety of additional jacks for plugging in external devices (see Figure 1-6). Some are inputs and some are outputs.

FIGURE 1-6   Typical input connectors

Input Connectors

Most camcorders have connectors for external microphones, usually mini-plugs for stereo microphones. Some are dual-function jacks, which accept external microphone signals as well as higher-powered signals from mixers, other camcorders, or VCRs. Better camcorders include three additional RCA jacks for video and audio inputs. These are usually color-coded red (right channel audio), white (left channel audio or monophonic), and yellow for composite video. DV camcorders usually also include S-Video (DIN) jacks for higher quality S-Video inputs. Some camcorders only have one set of jacks for input and output and determine which is in operation with a switch. A separate FireWire (I-Link or IEEE 1394) connector is also included for DV camcorders and usually functions as both input and output. FireWire ports carry both video and audio signals and other digital information, such as time code and titles. Some camcorders have external power jacks and other specialized connectors. Refer to your manual for these exceptions.

Output Connectors

Output connectors mirror the input connectors with the exception of the headphone jack, which is output only. As stated above, many camcorders supply only one set of jacks with a switch to change from input to output.

Become an Expert on Your Camcorder

1

It is important that you understand every aspect and feature of your camcorder. You should sit down and read your user manual from cover to cover. I carry mine in my camera case and refer to it often. Experiment with your camera and use every feature as described in the manual. Part of becoming a successful photographer means having the equipment become an extension of your body—you stop thinking of the camera as equipment and think of it as part of your vision. Finally, please remember that everything discussed in this book may be slightly different with your particular brand and model of camera.

Summary

Camcorders vary widely in features, nomenclature, and operating procedures. Check your instruction manual for details. The good news is that the quality of product is high and it is difficult to go wrong in product selection. Don't be afraid to use the trial and error process to learn how to operate your equipment. And have fun!

Chapter 2

Operate Your Digital Camcorder with Ease

How to...

- Hold the camera
- Plan your shots with editing in mind
- Understand basic camera shots and moves

Although digital camcorders put excellent video within the reach of even novice users, there are basic techniques that can make the difference between video that looks amateur and video that looks finished and well edited. This chapter will introduce you to the basic rules of shooting video, from holding the camera to basic shooting techniques, including the use of camera features such as zooming, picture composition, and continuity. You will begin learning about planning for editing and assembling sequences for interesting effects.

Holding the Camera Steady

You use a camera to capture your ideas on tape. To do this, the camera has to be in the right position *and* stable so that the objects being recorded are not blurry. Figure 2-1 shows how to hold the camera. Even if you are one of those people who can hold a camera incredibly still in both hands and are convinced you do it perfectly, I recommend you buy a tripod anyway. After all, no one—not even you—can hold a camera as steady as a tripod can.

By holding the camera in the classic position—in both hands, your eye looking into the viewfinder and your elbows resting at your sides—you can hold the camera relatively steady. Cameras with a pivot display are often easier to use since you don't have to brace the camera with your arms while looking through the viewfinder at the same time. If you are videotaping while walking, keep your knees relaxed to absorb shocks. When videotaping while standing still, use stable objects nearby, such as walls, cars, and poles, to keep your balance.

Everyone shakes. Even if you do not realize it, with an exposure time of 1/60th to1/50th of a second, the slightest movements are still discernable on videotape. If you do not have a camera that minimizes shaking and other vibrations with a built-in image stabilization system, your only choice is to use a tripod. In contrast to the electronic image stabilization system, a tripod does not influence the recording quality. However, if you still have to hold the camera by hand due to stylistic reasons or because of the particular subject being videotaped, imagine that your camera is a bowl of soup filled to the very edge and hold it accordingly.

Use monopods. These devices for holding a camera steady are similar to tripods, except they have only one leg instead of three. They are light and provide good support, yet offer more flexibility than a tripod in moving to a new position.

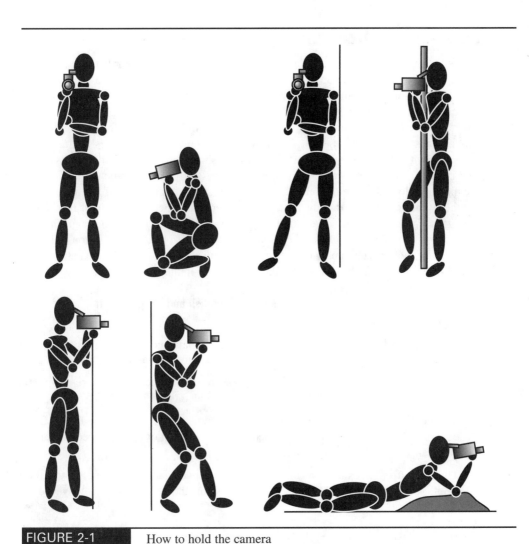

FIGURE 2-1   How to hold the camera

Basic Shooting Techniques

This section introduces the basic shots and camera moves you will need to master in order to take maximum advantage of your camcorder and create professional-looking video productions.

Shot Sizes and Zooming

Variety is important in all aspects of creating good video, and varying distance is no exception. A video shot entirely from one distance—that is, the distance between the camera and the subject never varies—would be unendurably dull. Physical and emotional reality, mood, and expression are established in good part by the distance. Take the example of a woman jumping a horse over a fence. A distant shot will have a totally different effect from a medium close-up shot of the same subject. In the first, you see the distant moving figures against lots of landscape with nothing very detailed. This type of shot sets the scene and establishes what the shot is about, the physical content, but does not necessarily provide much emotional impact. In the medium-close shot, on the other hand, perhaps shot from the end of the fence, you may see the sweat on the horse's body, the woman's hair flying, her expression of fear or elation as they lift over the fence and come thundering down on the other side. You are given much more information about both the physical and emotional content of the scene. If you followed these with a close shot of the woman's face, you would know by her expression whether she won or lost the race, and would be drawn into the story by the immediacy and intimacy of the close image.

By choosing the distance of your shot, you establish and control the impact of the video. There are four principal distance techniques:

The wide shot Establishes the subject and location of the action (see Figure 2-2). It is usually used at the beginning of a video or scene since it establishes the setting.

The medium shot Clarifies the subject or action, gives the viewer more information about the subject, and clarifies the theme of the video (see Figure 2-3).

The medium close-up Provides even more details for the viewer (see Figure 2-4). In this type of scene, the background has receded and the person or group of people being videotaped fill out the entire scene.

The close-up Zeros in on a single object—someone's face, for example—to fill out the whole scene (see Figure 2-5). Details of expression and feeling may be conveyed in a close-up shot.

One more word about close shots. They can be effective, even powerful, if used correctly. But this is often a case of "less is more." What may be interesting or compelling once or twice can rapidly become boring if overused.

Vary Your Camera Height

It's comfortable for most people to shoot video at their own eye level. It does not always produce the best results, however. A person sitting down will look better and be more comfortable if you videotape them from their eye level instead of your own if you are standing. This is especially important when videotaping children. A video that explores the world at their eye level will be much more effective and interesting. Modern cameras with pivot LCDs make it easier to watch what you are videotaping.

FIGURE 2-2   Wide shot

FIGURE 2-3   Medium shot

FIGURE 2-4   Medium close-up

FIGURE 2-5   Close-up

*The closer you are to an object producing sound, the stronger the sound will be.
Therefore, always be sure to check the sound level.*

Just as with distance, taping at different camera heights can suggest emotional effects or
mood. In general, it's best to choose one camera height to use throughout a video, except for
special effects. Extreme camera angles can provide a good effect if used well and sparingly.
Taping from ground level or from above a scene, for example, is an extreme angle and should
only be used if doing so clarifies the action or makes sense in the storyline. Varying the height
frequently within a video can be just as exhausting to watch as a jerky camera or too many
close-ups. It may be more tiring to shoot an entire children's party on your knees, but the
result will be worth it.

A "subjective" angle in videotaping means taping as if the camera's eye is that of one of
the characters. You are taping from the point of view of a specific person. This can be a
dramatic effect, but should be planned and used carefully.

Picture Composition

Picture composition is the breakdown of a picture into the foreground, middle ground, and
background. In all forms of photography and art, composition is critical in producing an
interesting and well-defined result. In spite of instinct, placing the subject always in the
dead center of the picture can be just that—dead boring. Good composition is usually
thought of in terms of an imaginary grid dividing up the picture plane. The *picture plane*
is the actual space in which the image exists—the paper, canvas, or photograph. In the
case of a video, the picture plane is the image as it appears on the screen of the monitor
or television. Imagine this picture plane divided into a grid of nine squares, three across
and three down (see Figure 2-6). These squares are sometimes referred to as *composition
zones*. In the center of the picture plane there are four intersections of these crisscrossing
lines of the grid. Placing the subject at any of these four intersections within the picture
plane ensures a good, balanced composition. If your subject is moving or cannot be placed
precisely where you wish, remember, there is always editing to come to the rescue!

This does not mean you can never center the subject of a video. In doing an interview, for
example, you want to catch every change of expression, and you usually want to see all of the
face. But a face does not always have to be dead center. Let the face look in toward the middle,
thus affording the picture space and movement. Experiment with placing the subject a little off
center, and see if it doesn't make a more interesting shot.

FIGURE 2-6   Nine composition zones

Focus Adjustments

You can focus most cameras in two ways: either manually or with the auto-focus function. Despite the leaps and bounds made in camera technology, you can't always count on auto-focus, especially when you really need it, such as for shots with a lot of contrast or when taping a moving background. Manual focusing still offers you the best control and gives you important stylistic advantages. Before taping, test whether manual focusing or auto-focusing provides better results. A small auxiliary monitor can be useful for this purpose. The one I use is a nine-inch portable television set with AV input connectors that can also be powered from my car battery. These are available from Sony, Panasonic, JVC, and others—all available at many consumer electronics stores.

 *The auto-focus of most cameras focuses on something in the middle of the picture. Therefore, if the subject you are videotaping is at the edge of the picture, be sure to focus in on it manually.*

Some situations demand manual focus. A videographer tells the story of videotaping the crater of the active volcano Stromboli in Italy. It was nighttime and he was waiting for the eruption. As he tried to let his camera auto-focus, it desperately attempted to focus in on something in the dark. When the volcano finally began sputtering, the auto-focus and automatic exposure went berserk and he ended up missing out on some great shots.

Videotaping Moving Objects: Panning and Tracking

Moving the camera horizontally while you are standing in a fixed position is called *panning*. Be careful to begin panning only when you know where you are going to start and finish—in other words, when you know what you want to achieve through panning. This type of camera movement is useful, for example, for following a moving object, scanning a landscape, or following the contours of an interesting architectural form.

> **TIP**   *Do not begin panning immediately after you switch on the camera, but instead, hold the camera for a few seconds at the start position, pan and then hold it again a few seconds at the end position. Try out the panning speed beforehand. Re-pan a shot if you think you may have shaken the camera since you can always edit out the unwanted footage later.*

Panning must be smooth if it is to be effective and look professional. When panning a moving object with your camera, it is usually best to use a tripod, ideally one with a fluid panning head that makes it easier to pan more steadily. A *fluid panning head* is a special top on the tripod that swivels and allows you to move the camera smoothly to follow the movement of your subject (see Figure 2-7).

Sometimes you will want to follow the action by actually going along with it. This is called *tracking*. Professionals sit on a camera truck called a dolly and have someone push them around on tracks to ensure they get a steady shot. This technique enables a videographer to shoot a moving subject while remaining beside it. You can achieve something of the same effect by practicing walking steadily with the camera in your hand. This is easiest if you are not looking through the eyepiece—that's why cameras with a built-in pivot display are so practical. With a bit of practice, you will be able to track relatively steadily.

Be sure to use the same size image area when tracking a moving object. If you keep this rule of thumb in mind, you can change positions. For instance, you can videotape a pedestrian from the side as you keep pace with him or her, then stop and let the pedestrian disappear out of the field of vision without running the danger of discontinuity.

Lines of Movement and Vision

When videotaping movement or action, you must pay attention to the lines of movement and vision. The *line of movement* means the actual path a moving object is taking. In other words, a person moving from point A to point B follows a line of movement. The *line of vision* refers to the distance directly between the eyes of two people who are looking at each other. Moving the camera angle from one side of either a line of movement or a line of vision to the other will look odd and confusing to the viewer. It is as if the viewer has been picked up and moved to the other side of the scene.

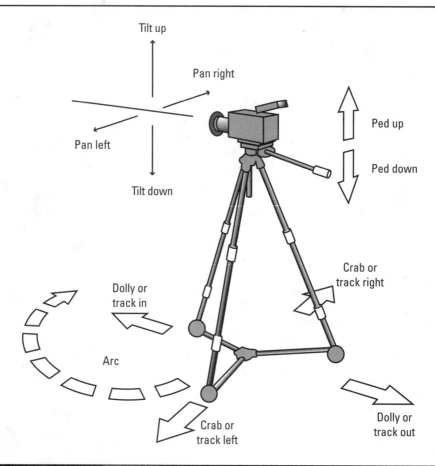

Tilt up

Pan right

Pan left

Tilt down

Ped up

Ped down

Crab or
track right

Dolly or
track in

Arc

Crab or
track left

Dolly or
track out

FIGURE 2-7   Panning head

Do not let the moving subject move too quickly to the outer edge of the picture since this causes the picture composition to appear imbalanced.

Panning eliminates the danger of crossing the line of movement since the viewer moves along with the camera and thus can easily comprehend the change in position. Panning is also easier if you use short focal lengths for taping. This means the zoom lens is completely screwed in, thus taping at a short distance from the subject.

If you own a tripod, you can put it on a platform with castors (just make sure it cannot fall off). If you don't have a tripod, you can use a shopping cart, a baby carriage, or an open, slow-moving convertible.

Continuity

If you are taping a story or making a video where a sequence of events is important, keep your continuity in mind. *Continuity* means that a series of scenes make sense, follow one another logically, and don't contradict in later scenes what was established in an earlier one. For example, in one scene of *The Maltese Falcon*, Humphrey Bogart managed to burn his cigarette down and then have it grow again. It is clear that the person responsible for continuity was not paying attention and different scenes filmed at different times were accidentally pasted together. A more modern example appeared in the Kevin Costner version of *Robin Hood*. The evil Guy of Gisborne was established in an early scene as having a long scar on one side of his face. In a subsequent scene this scar had mysteriously migrated to the other side. While few of us engage in creating epic (or attempted epic) films, you may want to try your hand at videotaping a story where sequence and continuity are important. A written list or script can be a big help in managing continuity by keeping track of both the logical visual and chronological sequence in a video.

Another issue of continuity involves changes of light and shadow when you tape over several hours or on different days. As you know, light changes during the course of a day, and the camera is sensitive to light and shadow. Tape of a scene shot at ten in the morning will look quite noticeably different from the same scene shot at two in the afternoon. Both the color and quantity of light changes, as well as color and placement of shadows. If the light changes, postpone taping until the next day or retake all shots with the new light situation. Severe light changes are very obvious to viewers.

Planning and Editing

To avoid any unpleasant surprises, carefully plan how you are going to shoot your video. Have a good project concept, and consider putting it down on paper before videotaping. Even in an informal situation, such as a child's birthday party, it is a good idea to have a plan of action in mind. Make note of special activities, where the action will take place, how many people will be involved, and what particular shots you want to get.

If you are going to tape, as they say, "on location," in some cases you may need a permit or permission. Some churches, for example, have rules limiting the use of cameras and video equipment during services and weddings. If you are planning to make a video of a wedding, check to see if permission must be secured, and decide on the best location for the camera. All this should be done well in advance of the event!

When deciding on different shots and camera positions, remember that they will later have to be put together to create a harmonious scene. It does not make any sense to tape different shot sizes that do not fit together. They will only cause you trouble later at your editing desk.

For every action in a video script, professional camera people create an appropriate shot. If you are using your camera to shoot a family reunion or birthday party, you do not have to be quite as exact. For the more advanced projects you tackle, it's smart to give some thought ahead of time to the various shots required for each scene. Production planning and script writing are covered in more depth in Chapters 16 through 19.

Remember, it is always better to have too much material than too little. Professional video producers speak of the "shooting ratio," which is the ratio of tape shot to tape used in the final production. This ratio can be 20 to 1 in many cases. If you shoot at 2 to 1 (twice the footage you finally use), you may consider yourself organized and efficient. Remember, tape is cheap, and the video situation may not ever be recreated for a second chance. When in doubt, shoot more!

Assembling Sequences for Interesting Effects

Editing is essentially taking a lot of bits of video and creating a deliberate, logical, and effective sequence—usually with a beginning, middle, and end. It is not always possible to shoot all the video needed for a project on one day or in one location. Editing can put it all together as though everything transpired in the order you present in the finished video. In almost all the movies you see in theaters, various scenes were shot at different times and in almost any order. (It's one of the complaints of actors that it is difficult to maintain an emotional mood when the sequence of scenes shot does not reflect the sequence of the story.) Editing takes this apparently disassociated jumble of scenes and creates an artful telling of a story. It does this not only by arranging the scenes in a sequence that follows the story's events, but also by juxtaposing images and suggesting relationships. Many a great film owes much of its greatness to the film editor.

Suppose you are videotaping a woman entering a house. She goes up the steps, searches for her key, and unlocks the door. As she pushes the door open, you cease taping. You want the woman to enter a particular hallway, and it doesn't happen to be the hallway on the other side of the door you are taping from the outside. So, in a separate sequence, you tape the woman entering the hallway—in this case, the hallway you want to appear in the video. Provided the woman is dressed the same, you can edit the two pieces of video together for a seamless bit of storytelling. The viewer will have no reason to stop and think about whether the original outside shot is the same building as the one the woman enters in the inside shot. This is referred to as *location logic*.

Another type of creative sequencing of shots through editing is that of *causality logic*. Suppose you videotape a handsome man in a garden talking to a pretty woman coquettishly twirling a parasol and fluttering her eyelashes. You show a close-up of the man's face, then you edit in a quick shot of a small garden snake slithering away into the bushes. We now have a very particular understanding of the man, even though there is nothing in the original garden scene to suggest his character. If you edit the snake shot into the video following a close-up of the woman, you have a different understanding. Two disparate pieces of video have been, in each case, edited together to convey a message or make a point by their juxtaposition.

Now, you're probably asking, "What does that have to do with me?" It means that even if you never make a movie, you can use the process of editing to create videos that are not only adequate but artful. Random footage of a birthday party is never going to be as fun or interesting to watch as a video of the event that's been edited together with an eye to content and sequence.

Last but not least, here are the six pitfalls of shooting video. Avoid them at all costs!

Snap shooting   Avoid recording for less than four seconds. This makes your video choppy and very difficult to edit later on.

Constant zooming   Do not zoom too much. Zooming is interesting once in a while, when it actually contributes to the telling of the story, or is dramatically good at a certain point. Overused, it is distracting and hard to watch.

Garden-hosing   Avoid restless panning without any defined purpose. This simply looks amateurish and also makes the video hard to watch.

Stubborn taping I   Do not always tape from the same height. Variety is the spice of videotaping, provided it is not overdone. Varying the height of the camcorder when appropriate can add drama and panache to the videotape. Don't, however, overdo this.

Stubborn taping II   Do not always tape from the same position. As with the height of the camcorder, the position of the camcorder relative to the subject should vary as the action or mood of the video suggests.

Blinding effects   Avoid having the object's background lighter than the foreground being taped. Your object in creating videotape is not to make the viewer squint, trying to make out the subject. Unless you are deliberately trying to obscure the foreground for some reason, make sure the lighting is appropriately on the areas of interest.

Summary

This chapter described a variety of shooting techniques—from how to hold the camcorder to how to use such basic functions as zooming and panning—as well as the importance of composition to give you a kick-start to becoming an award-winning videographer. Planning at the beginning of the project was also suggested so that you end up with the shots you want and have enough quality material to edit later. Don't forget to review Chapters 4 and 5 to learn about the important aspects of lighting and sound.

Chapter 3

Shoot Effective Video of a Variety of Subjects

How to...

■ Shoot video of people, children, animals, and sporting events

■ Shoot video of landscapes and architecture

■ Get the best results when you videotape interviews

You've got your camera in hand and are dying to start shooting. If you have a lot of experience, you may have a relatively good idea of which shots are particularly important. However, if you are a beginner, you should become familiar with the material you propose to shoot. There are, of course, an infinite variety of subjects to videotape. Although many of the basic techniques for shooting good video are pretty much the same regardless of subject, there are some tips specific to the category of subject you choose. One thing is true no matter what your subject is: as the videographer you can either produce a collection of nice but unrelated shots or an impressive video with a lot of thought behind it.

If you want to learn to create the latter, you'll find this chapter very handy. It is not my intent to provide you with a high-level or advanced how-to guide, but rather to inspire you with these tips to find out for yourself what works best for you. This information will also allow you to be more critical of your own work and that of others.

■ Where will you be videotaping?

■ When will you be videotaping?

■ Who will be in your video?

■ How do you want this video to turn out in terms of style?

■ Why is one scene or another important to the end result?

Use this basic information as a point of reference for your project. The more points of reference you have, the easier the videotaping process will be. With each subsequent video project, this preparatory work of creating a concept will become easier.

Videotaping People

People are particularly exciting to tape. Taping people is always a challenge because you have to keep many things in mind. Facial expressions, gestures of hands and body, posture, and position all give clues to a person's mood, character, and situation. Don't just be a passive cameraperson when you're taping people. Interactive communication with your subjects is essential if you are to capture their uniqueness on videotape. The quality of your video and the degree of personal revelation it will have will be determined by how sensitive you, as the cameraperson, are to your subjects and their particular situations. Before you begin taping, decide which of the following aspects of the people being videotaped should be emphasized:

Character   It has been said that the camera doesn't lie. With modern-day effects packages and editing, this may no longer be a true statement. Nevertheless, video can be revealing of a person's character. Taping a person rolling around with a gaggle of children and puppies, laughing, tells the viewer something—though by no means, everything—about them. The same person sitting at a desk, absorbed in work, bending over a sickbed, or digging in the garden shows different facets of their character to the camera.

Special talents   Is the person a potter, a painter, a carpenter? Taping someone engaged in his or her special work is a powerful way to convey the essence of a person.

Age   We live in an age-conscious world. Sometimes the age of the person in the video is not of primary importance—if you are focusing on their work, for example. Other times, age may be the point, such as in a video about nursing homes or community programs for the elderly. Camera angles, degree of close-up, and lighting all contribute to how important age is to a video and how clearly it is revealed.

Current personal situation   Is the person happy? Sad? Angry? Triumphant? Have they recently been promoted or fired? Are they in good health or facing serious illness?

Location   Where are you taping the people in your video, and is the location important or incidental to the video? Picture a small child, alone and crying. She is standing in her backyard. Now picture the same child, alone and crying, but the location is Disneyland. Just by showing location, the story is a very different one.

The time and time span in which the story is taking place   Are you taping a single event, such as a party or a graduation? Or are you making a video that chronicles a week of someone's life?

TIP   *When taping people, good quality sound is of the greatest importance. Sometimes you will be adding the sound after the videotape is shot, using narration, for example. If it is important to record the original sound when taping and if your camera's microphone is not very good, use an external microphone that can be operated by an assistant.*

Planning the Best Sequence of Shots

Keep in mind what you want to emphasize in your video, and plan a sequence of shots that will convey your ideas to the viewer. A generic sequence of shots could go something like this:

1.   Introduce your characters to the audience. Use a wide shot (showing the scene location and all the characters) to capture the entire picture so the audience can orient itself.

2. Do not hold the wide shot too long. Wide shots are designed to give a lot of information in a scene. As there is not much detail, they will not hold interest for long. A medium shot moves closer to the subjects, the background recedes, and the theme or action of the scene is clarified by a closer, more detailed look. You will need to move in to a medium shot to provide more information about the people in the scene. An exception to this is to use a tracking shot to investigate an area before you move in to a medium shot.

3. Tape each individual the same length of time. This way you can choose the taping rhythm that suits you and decide later during editing where you want to place the emphasis. When you edit the video, you can shorten shots of some individuals who are not the main characters of your video, while using the full amount of time focused on others who are essential to the sequence or story. If you start with the same amount of tape for everyone in a scene, you have the maximum flexibility in how you edit it together. The point here is not watching the clock, but taping more than you need of everyone so you have plenty to prune from. You can always cut away what is not interesting, but you can't often go back and get more to paste in. (Obviously, there are times when you are interviewing someone when the content dictates the limit of time taping.)

4. Study the setting and integrate the people you are portraying in it. A group of people, for example, is gathered in a sidewalk café. There is a romantically involved couple at one table and a long-married couple at another. Other people in the café include two businessmen, a pretty single girl, and a harried mother with three rambunctious children. Compose the shots so they capture the relationship of the individuals to each other and their environment.

For action shots, change positions often to provide more perspectives of the environment and the people being taped. Scenes with a lot of action can usually be effectively conveyed to the viewer only by changing angles and image sizes. Perhaps you are taping a basketball game. A distance shot shows the court, the players running up and down the court, and some background of the viewing fans. This distance shot will not convey nearly the amount of action as one taken from courtside, and still less than one taken from immediately behind a basket as the players thunder toward the camera. A close-up of one player's face leaping toward the basket conveys the effort and determination of the players. To achieve all this later when editing, you have to have taped the right material to begin with, so plan ahead about the different shots, camera positions, and angles you need to use.

 *Long, important shots of a person's face should always be made with a tripod to ensure high quality. In contrast, tripods are useless when working with quick position changes.*

As discussed earlier, vary the shots that you shoot. However, avoid deadly sins like crossing the line of movement or taping from only one position. As the cameraperson you should be actively involved with the action.

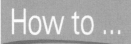
How to ...

Get the Best Results When You Videotape People

■ Decide in advance what personal aspects of the subjects you want to emphasize.

■ Introduce the subjects to the viewer (for example, from wide shot to medium shots to close-ups).

■ Use fast pans for fast action.

■ Change positions, but be careful not to cross the line of movement/vision.

■ Record the sound separately.

General Guidelines for Videotaping Children

Children are generally videotaped with an emphasis on fun. Since children know what a camera is, you can choose between two approaches: keep the camera as inconspicuous as possible in the background to ensure spontaneous and unrehearsed shots, or let the children consciously react to the camera. In this case, it should be clear that many of their reactions will be provoked by the presence of the camera.

Tape children at their eye level (see Figure 3-1). This creates significantly less distance between them and the viewer.

Always be prepared for surprises when taping children. You may have to change the defined shooting schedule on short notice to handle another subject that is perhaps more appropriate at a given moment. Children are spontaneous, but also usually easy to motivate. Try interviewing them—some may end up telling more than you expect. Let kids act a little. Some children love giving a small performance and enjoy the attention.

Protect your camera when working with children. A quick solution in a pinch is to cut holes in a plastic bag and put it over your camera. This way you don't have to worry about flying cake and water from squirt guns ending up in your camera.

Here are some general guidelines for taping children:

■ Is what you plan to tape somewhat unpredictable, such as a children's birthday party? If so, make a rough outline of the situation. For example, who is the birthday child, how many children will be at the table later, and what games will be played?

FIGURE 3-1   Taping children at their eye level

- Begin taping with a wide shot. Tape the group of children as a whole and then begin making portraits of individual faces. Change shot sizes depending on the action taking place. Make things easier for yourself by not using the zoom too extensively.

- Be sure to tape close-ups. Focus on the individual children's faces, moving hands, dancing feet, birthday cards, and other details. This provides valuable material for cutaways. These are shots that jump from the subject to something that amplifies or explains a situation—a cutaway from a child's delighted face to a pile of wrapped packages, for example.

- Change observation points and pay attention to the actions and reactions of the main child involved. If a child is involved with a toy, for example, use a reverse shot. Focus on the toy and then back on the child's face.

- If another person (another child or an adult) begins participating in the action, tape this from a different angle. You may choose to tape from their subjective angle (their eye level and relative position).

- End the sequence with a pleasant event such as a bunch of colorful balloons with close-ups of faces and hands.

3

TIP *Adjust the microphone so that it does not over-modulate when you are talking to children with the camera in your hand. It is best to try it out beforehand.*

Get Great Results When You Videotape Children

■ Always position the camera out of sight if you want the children's behavior to be spontaneous and candid.

■ Start with a wide shot and then switch to medium shots and close-ups.

■ Shoot from children's eye level.

■ Emphasize the subject of the video. The birthday child's family will be watching the video, and they will be most interested in his/her reactions.

■ Frequently change observation points while remembering not to cross the line of vision.

Videotaping Animals

There is a big difference between taping wild animals and house pets. Taping wild animals requires much more time, effort, and materials, and occasionally (but not always) more courage.

We may not all be ready to go on safari and tape animals in the jungle, but we've all seen home movies of the bears in Yellowstone or bison in fields near the roadside. When taping animals in nature, it's a good idea to collect extensive information about the kind of animal you plan to tape so you can predict its behavior when taping. Make a checklist of its eating habits (time and place), habitat (in forest or open terrain), and shelters (size and how often they go there). Good places to get information are regional animal associations, rangers, universities, and your local library. The more you know about the animal you plan to tape, the fewer difficulties you will encounter during the shoot.

Be sure to keep the following in mind when taping animals:

■ To better familiarize the audience with the animal, provide a few shots of its habitat. Use a wide or medium shot or even a tracking shot to sufficiently describe the location.

■ Adapt the shot size to the animal's height. For example, the setting for a dachshund in the park will be lower than that of a zebra in the savanna.

■ Look for a good location to shoot from, one that minimizes distractions (too many people in the way, for example) and gives you a good view of the animal (see Figure 3-2). If you want to distinguish the animals from the background, you may want to set your camera to manual focus and try to make the background out of focus. If this does not work and the background is visually disturbing due to people, walls, fences, and such, use a different shot.

FIGURE 3-2   Framing shots of animals

■ Give animals time to get used to you. Accept the fact that it will probably take some time before you get any interesting shots.

■ When taping pets, try taking cutaways from the animal's perspective (subjective angle).

■ You should have a basic storyline in mind even when taping animals. Find out more about the animal you are taping and where the action is taking place so you can think up a good story.

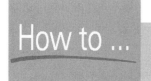

Wait for activities animals perform every day, such as eating. This is an easy way to get good action shots. In zoos you can sometimes talk to animal caretakers and get tips from them. You may also be allowed to shoot from particularly advantageous positions.

How to ...   Get Great Results When You Videotape Animals

- Gather information on the animal to be taped. Integrate this information in your storyline.
- Look for good shooting locations.
- Add reverse shots from the animal's perspective/eye level.
- Avoid distracting backgrounds.

Videotaping Sports

Documenting a sporting event requires a lot of preparation, since sports thrive on highlights and fast action. In taping sports, keep the line of movement in mind. If you constantly change positions—for example, from one side of the field to another—your viewer will be confused. When editing for highlights, also capture the reaction of the fans. This creates the mood of the event for your viewers.

Choose a clear action sequence as your goal. At competitions, try to observe the athletes while they are getting ready. Then follow the athletes with the camera to the actual field or court. Get close-ups of the players. Shooting a player moving directly toward you can also be effective, such as shooting from the turn at a car race.

If possible, incorporate interviews into your video. They can add tremendous interest.

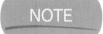 *Bring along enough batteries or make sure you have access to an electrical outlet. If it is cold, keep the batteries warm—in an insulated bag, for example—so they do not discharge as quickly.*

When taping family sporting activities such as skiing, you have a less complex environment than a busy stadium during a sporting event. You will need to analyze the line of movement, which is not difficult with skiers or snowboarders, especially on a straight downhill slope. For variety and to heighten the sense of action, quickly change to other spots for watching the action. You may want to note how many shots you have taken from a particular spot.

Pay attention to the development of the action sequence. Use the same rules described in the section "Videotaping People," that is, move from wide shots to medium shots.

Do not use a tripod when taping action sports, since it limits your range of movement and reaction. Hold the camera by hand, making sure you are holding the camera steady.

TIP *To make working with a monopod quicker and easier, modify it a bit: screw a round metal plate to the foot of the monopod. It should weigh at least one pound (the lighter the camera, the less weight needed). This lets you hold the monopod and camera as if you have one of those sophisticated camera stabilizers. The additional weight makes it easier to hold the camera steady.*

Be sure to use the right shutter speed. Fast-moving bodies can be viewed and analyzed later in slow motion. However, if the shutter speed is too low, they will appear blurred. Take a few test shots to find the right speed.

Remember to incorporate general shots to give the viewer an overview of what's going on. The color and movement of spectators, fans, and scenery can all add excitement and interest. They can also be used later in edit as cutaways.

In every video project, it's a good idea to identify your video by noting the name of the subject and date of recording. This is important background information for the viewer. The more information you gather when shooting, the better—it is always more difficult to track down information later.

TIP *When creating tracking shots freehand, be sure to protect your camera. You can buy protective carriers, but a little ingenuity can produce one that is inexpensive and works very well. Take an empty plastic jug (for example, the kind distilled water comes in, not oil). Cut the jug right under the handle, flip it open, and cut a hole about the size of the lens. Place lint-free foam rubber in the jug so that the camera fits snugly inside. Close the "action box" with gaffer tape. With your camera in the action box, you can race behind skiers and skateboarders, and if you have an accident your camera will be relatively well protected.*

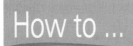

Get Great Results When You Videotape Sporting Events

- ■ Define and maintain the line of movement, and do not cross the line.
- ■ Analyze the action sequence and define camera positions.
- ■ Tape close-ups showing faces and details such as numbers on jerseys, hands reaching for the ball, and so on.
- ■ Use cutaways, especially at highlights, with reactions of spectators.
- ■ If necessary, use faster shutter speeds for action shots.
- ■ Pay attention to light situations, especially at outdoor sporting events where lighting can change suddenly.
- ■ Protect your camera.

Capturing Landscapes and Architecture on Video

Landscapes and architectural elements provide their own challenges to the videographer. The details the eye can see get lost when you shoot wide. Television is called the "medium of the big close-up" because that is what it does very well. Landscape works on a 30-foot film screen, but it does not work as well on a 32-inch TV screen.

Unlike videos of people, action is not the most important focus of landscape and architectural video, although there may be action within it. Trains pass through deep valleys, birds fly behind or over buildings, and water rushes beneath a bridge. Yet the purpose of the video may be not the motion, but the landscape or the structure itself.

Guidelines for Videotaping Landscapes

Landscapes contain both horizontal and vertical lines (see Figure 3-3). Scenes shot from a distance or scenes of fields, for example, will be largely horizontal. In scenes of mountains, tall trees, or waterfalls, vertical lines dominate. Regardless of whether you create your composition from vertical or horizontal lines, train your eye to spot these lines in your subject. Set the correct, most effective horizontal line. Decide whether the sky or ground should dominate. This can be important, for example, for conveying the mood of a storm. In this case, the horizontal line should be relatively low and the picture dominated by sky and clouds. A garden full of flowers, on the other hand, would focus attention toward the ground and the sky would be less important, so the horizon line would be high. In many horizontal compositions, a vertical element adds interest and depth. A view of a valley with mountain ranges in the distance may have more impact if there is a tree in the foreground and to one side of the picture plane. A vertical composition may be too static and boring if the subject is placed in the direct center of the picture. A waterfall or a majestic redwood can still be the "center of attention" though off-center in the picture plane.

FIGURE 3-3   Typical landscape framing

In shooting landscapes, you must learn to define the perspective. The viewer should be able to recognize and feel depth and distance in your pictures. It is important to find the right shot to illustrate a landscape's size and uniqueness. A successful landscape composition often includes people to give the viewer an idea of scale and perspective.

Light and color can change quickly when you're shooting outdoors. Be sure you are familiar with local light conditions. Observe the change in colors of your subject. Tape as early in the morning as possible when the light is best. At noon, pictures appear flat due to a lack of shadows.

A bright blue sky may look pretty, but moving clouds often make for more interesting shots (fast clouds = fast change between light and shadow). They appear dramatic and can be used during editing to create a slow-motion effect to intensify the overall presentation of the subject matter.

When it comes to landscape shots, people often zoom too much. It is better to have good shot sizes that you can later use during editing to create the right atmosphere and dimensions.

3

 For tracking shots using a car, make sure the camera is directed at the subject (a shot of the dashboard is not that impressive). Try to dampen vibrations as much as possible. This can be done in two ways:

- *Hold the camera in your hand (you can also use a modified monopod). Make sure you can move your arms freely to continuously balance the camera.*
- *Use a suction cup attached to the hood or windshield of your car to mount your camera. This, of course, does not reduce vibrations quite as well.*

General Guidelines for Videotaping Architecture

In architecture shots, vertical lines often dominate (see Figure 3-4). The camera can be used to emphasize or lessen these lines. Through skillful taping and editing, you can present the same architectural element in many different ways, such as emphasizing the visual beauty or the ugliness of a structure. Note special architectonic features and shoot these details with a telephoto lens using a tripod. Learn as much as possible about the structure beforehand so that you recognize perspectives and details that you might have otherwise overlooked.

Be sure to use a tripod or other secure base when shooting architectural structures. Remember that the structure you are shooting is always steadier than your hands, so every little jerk of the camera will be noticed.

When shooting during the day, use the automatic exposure. At night, however, automatic exposure can be a problem. Test it out in your particular situation. Perform a trial pan to spot changes in lighting conditions. Lighting can vary greatly, especially with large structures (lighting varies from top to bottom on a structure as you move away from the sky).

An unprepared videographer can easily be overwhelmed by the onslaught of details in a large city. To avoid this problem, define exactly what you want to capture and how you want to create a dramatic interaction between these details. This is best done by drawing up a shooting plan in which you specify which shots—wide shots, medium shots, with and without people—you want to make of various buildings.

Just as you would with a people-oriented video, develop a storyline for your architectural video. You may want to begin or end your video journey with a close-up of an architectonic highlight.

FIGURE 3-4   Framing architectural shots

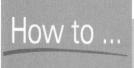 **Get Great Results When You Videotape Landscapes and Architecture**

- Define and utilize dominant geometric features.
- Assess light and shadow.
- Define perspective.
- Use a tripod whenever possible.
- Perform trial pans.
- Do not shoot large buildings with automatic exposure.

Getting the Best Interviews on Videotape

It's a great idea to use interviews in your video. They provide information, present opinions, and help the viewer become more familiar with the characters in your video. There are a number of projects you might undertake that could involve interviews. Taping a family member talking about their childhood memories can provide a priceless record for family history. People in a church or organization may be interviewed to give historical background and discuss changes in the organization over time. Interviews with company managers and employees can bring company videos to life. And, of course, you could make "person in the street" interviews on almost any subject.

When speech is vital to the finished video, as in an interview, always work with a separate microphone if you have that option. The sound quality delivered by the average video camera is really not good enough for projects involving interviews. If you do not have a separate microphone, buy one. You'll definitely be happier with the better sound it delivers.

Inform the person you are interviewing what questions you are going to ask, but do not discuss the questions with them in detail. This gives the interviewee the opportunity to prepare for the topic, but also lets you interject things spontaneously. Exceptions to this are quick, informal interviews, such as comments gathered at sporting events. However, at this kind of interview, the person still knows what the interview is about.

When interviewing specialists in some particular field, be sure to thoroughly familiarize yourself with the subject beforehand. If the person just won a race, know whether this is the first race they have won or part of a winning streak. If you are interviewing a politician, know what their platform is. If you are interviewing a writer, read some of their work. This kind of background research will enable you to prepare thoughtful, relevant questions. Prepare a list of key points so you can phrase your statements and questions adeptly and to ensure you really address the most important and interesting points.

For a formal interview, make sure you have the right background at your shooting location (see Figure 3-5). An overly busy background will detract from the subject. Consider where you want to conduct the interview and how it will be later integrated into the rest of the video. Check whether the conditions at this location—such as light, noise, and weather—meet your expectations. Remember that you can conduct interviews much better in a pleasant environment.

Work with assistants when taping formal, prepared interviews. Let them handle the camera, sound, and lights. It is essential that the subject of the interview feels comfortable. It is also important that you are well organized and not distracted by trying to pay attention to too many technical details.

When taping an interview, keep the following in mind:

■ There are several ways to begin an interview. You could start with a wide shot of all participants and ask your first question during this sequence. Or you could shoot the subject walking, playing, or doing whatever it is they do, and then cut to the interview.

FIGURE 3-5   Framing an interview shot

- After taping with a wide shot for a few seconds, reduce to a medium shot to focus in on the action in the foreground. Stop taping and ask the interviewee to not reply to the question yet, so the video does not show the transition between the medium and wide shots. Continue taping after you have zoomed in on the person from the wide shot. Have the interviewee answer the question. Of course, be sure to tell the people being interviewed about this technique in advance!

- Generally the cameraperson can, at his/her own discretion, decide when to use the above procedure to zoom in and out on the person being interviewed. They should be guided by the subject's gestures and mood changes.

- Sometimes it is also good to take a reverse shot of yourself, the interviewer. If necessary, ask the question again.

- Listen to your subject. Your prepared questions are a guide to keep you on track, they are not written in stone. Be prepared to go with interesting off-subject topics.

- Be careful to not step on your subject's answer. Wait a beat before asking the next question so you have a discernable end of the comment in order to trim out dead space when you edit. This is called the "out point" in editing.

- In general, shoot interviews as close as possible, using close-up or extreme close-up shots.

- Locate the interviewer slightly off camera and have the subject talk to the interviewer, not the camera. The subject will be more comfortable, and the interviewer can make the experience seem more intimate.

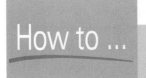 **Get the Best Interviews on Videotape**

- Prepare for the subject of interview. Know your subject!
- Have a preliminary talk with interview participants about topics to be covered, your shooting technique, pauses, and so on.
- Pick a location for the interview based on suitability of light, sound, and environment.
- Use an external microphone and a tripod.
- Use assistants.
- Start with a wide shot, then change to close-ups of interviewees.
- Change shot size, keeping in mind the rhythm in which questions are asked.

Spontaneous Videotaping

To tape something spontaneously does not mean to simply push the Start button and record whatever happens, but rather to capture something that was not planned much in advance. This could be a spontaneous family outing or your child's first steps. The following are a few tips that can help you get the best results:

- Change positions frequently so that you have a variety of material to choose from when later assembling your video. Obviously, if you are getting a great shot, stay with it until you have everything you need before moving on. Have a firm stance and sufficient freedom of movement so that you do not endanger yourself or others. When panning without a tripod, rotate from the hips. Be sure to breathe easily and stay relaxed.

■ Heed this basic rule of acting, which also applies to taping: be sure to always carry out a movement to its completion. In other words, do not let your movements be interrupted. Always define a beginning and end to your pans and camera movements and rigorously follow through on them.

■ Do not switch off your camera between quick successive actions. Instead, move the camera quickly to the next object. It is easier to remove the camera movements later during editing than to perhaps miss out on some important and interesting shots.

■ Do not zoom too much. Only use the zoom function to emphasize the perspective of a location or to change to another shot size. Excessive zooms are tiring for the viewer. Once you have achieved the ideal focal length, stay at this setting for a while.

■ Do not shoot any takes less than four seconds long. The results will be choppy, and it is a real pain during editing to deal with tiny snips. Be sure to roll the camera before the shot and continue to roll after completion of the shot so you have extra film, or "head and tail," for edit. If you don't include this extra tape between shots, it makes it impossible to edit without losing good tape.

■ Learn from your mistakes. Examine how you taped spontaneous, unpredictable situations in the past. Look at your old material and decide what you would do differently next time.

■ Experiment! Just like your signature, your video should reflect a very personal side of you. Use your imagination and dare to try new things.

Summary

Now you are all set to start planning the subjects of your upcoming video projects. Shooting good video involves many of the same basic techniques no matter what the subject is, and planning in advance regarding location, content, and style will take you a long way toward a quality product when you are finished. People, animals, landscape, and architecture all present their own special challenges, but all of these categories can make fascinating footage. The key principle is to think about what you are doing, analyze and study the subject so that you understand its properties, its strengths and weaknesses, and then work with the strengths. Finally, remember that the finished video on the screen is the objective. Always make the shot that will, in your mind's eye, be interesting on the screen.

Chapter 4

Get the Best Lighting for Your Videos

How to...

- Understand the basics of lighting
- Design basic video lighting
- Select proper lighting instruments
- Understand the role of color in lighting
- Solve lighting problems with basic techniques
- Understand the impact of lighting on the operation of your digital camcorder

Although light and its effects are all around us all the time, the study of lighting as a craft is a highly developed one. The basics of lighting design, however, are easily mastered. Careful observation of the effects of lighting on subjects and the resulting video is a great way to learn. There are, of course, fine books and resources available on the subject of lighting for video production. Remember, the best way to learn is through experience—moving the lights around yourself, making adjustments, and seeing the results. Working with an expert, if you can, is a great way to learn lighting tips and tricks. This chapter offers a look at lighting concepts and techniques, and addresses issues of particular interest for home digital video.

The Basics of Lighting

Good lighting design comes down to basic questions such as how bright the available light is, where the light is coming from, and whether the light is warm or cool in color. In a short time you will be able to judge these factors of light and make them work for you in designing your video. The main issues in understanding lighting for video are:

Intensity   This refers to the brightness of the light. Light intensity affects exposure.

Quality   This refers to whether the light is a point (single, sharp) source creating hard shadows, or a diffuse (soft, with no sharp edges) source creating soft shadows with less light.

Contrast   This refers to the relative brightness of the darkest to the lightest areas in a shot.

Direction   This is the light source relative to the camera's position. This affects the appearance of the subjects.

Color temperature   This is the spectrum of the light or the mix of colors present in the light and their relative intensities. This affects the overall color relationships in the shot.

This may sound complicated, but to shoot quality video you must understand how to use light to its best advantage for your finished product.

Light Intensity

Light intensity simply means how bright the light is. A video camera CCD chip requires a certain amount of light to operate properly. Contemporary digital cameras have improved a great deal over earlier technology and can function with much lower light levels. There is, however, a threshold of quality as well as function, and a good level of light is a fundamental requirement. The issues that affect light quantity for the camera include the following.

The Amount of Illumination Falling on the Subject

4

You may not always want bright light on the subject, but you will always want the amount of light necessary to help create the mood and look you are after. Just as important, you want there to be enough light so your video camera works at its best. This may be adjusted by increasing or decreasing the light source's intensity—that is, making the light brighter or less bright, or moving the subject closer to the source of light. You can also effect light quality by your choice of a focusable light versus a fixed or unfocusable light. Focusable light gives you the ability to change the light's output level by an adjustment of the lamp, iris (an adjustable device made up of overlapping metal plates that opens and closes to adjust the amount of light), or barn doors (adjustable shutters over the lamp of a light fixture). A fixed or unfocusable light must be physically moved.

Lens Aperture or F-stop Settings

This small aperture on your camera controls the amount of light reaching the CCD chip inside. No matter how much light you have on your subject, you won't get the effect you want if the f-stop is incorrectly set. Adjusting the aperture increases or decreases the amount of light. Be aware that increasing the aperture decreases the depth of field—in other words, making the aperture larger, which lets in more light, also reduces the area of sharp focus.

Filters

Filters also reduce the amount of light transmitted to the lens. The decrease in light transmission caused by a filter may be compensated for by increasing the aperture or the light quantity. The number of f-stops that the filter reduces the light is usually indicated on the filter's package or documentation.

The best way to determine the light quantity with consumer cameras is to simply look through the viewfinder. The best teacher when it comes to lighting is to pay attention to the quality and results of lighting installation and to the results on the video camera or film. The photographer's eye is an invaluable tool.

Quality of Light

It may be helpful to think about lighting as having less to do with light itself, and more to do with controlling shadows, where the shadows are, and what type of shadows they are. Most people pay little attention to lighting that adequately illuminates its subject, but will notice when the subject is underlit or if the lighting casts distracting shadows.

There are two basic types of light in the natural world, sunlight and overcast light. All artificial lighting sources and styles mimic these two fundamental types of light. Sunlight is direct light. It comes from a single, identifiable source and creates clearly defined shadows. Look outside on a bright sunny day, and notice how trees and buildings throw sharp, deep shadows. Sunlight tends to be harsher than overcast light, but also more effectively shows the texture and shape of the subject.

Overcast light is indirect light. It is diffused or reflected, coming from a larger source and casting softer shadows. It is flatter, more two-dimensional, but can smooth out a scene and make it more pleasing to the eye. Shooting video indoors, especially for anything requiring setup lighting, is obviously more complicated than using the sun as your only illumination. When shooting video for a project requiring more than just available light, you will discover that a good lighting setup is usually achieved using a combination of light sources and adjustment techniques that works best for a given subject or effect.

Use Direct Source or Hard Lighting to Get Distinct Shadows

Direct or hard lighting is more controlled and results in hard shadows. Sources of direct light include the sun, as well as artificial lights with lenses, bare lamps, or smooth reflectors (those smooth metal cups that partially enclose a lamp and reflect the light outward).

Positive attributes of hard lighting include:

- Clear and sharp shadows, often producing dramatic effects.

- It is directional and easy to control, but requires accessories such as lenses.

Negative attributes of hard lighting include:

- Harsh or distracting shadows in some cases, particularly on backgrounds.

- It provides narrow coverage requiring more lighting sources.

- It creates multiple shadows resulting from multiple hard sources.

Use Indirect/Diffuse or Soft Lighting for Softer Shadows

Diffused lighting is a soft light that produces few or soft shadows. It is more even than direct lighting, but produces little contrast or modeling on subjects. In other words, the subject may appear flatter, with less sense of three dimensions and less detail. Soft light is much more natural in our environment. It is rare to use hard directional light other than direct sunlight.

Positive attributes of soft lighting include:

■ It produces subtle shading.

■ It reduces modeling and textures when you want a smoother effect.

■ It doesn't produce unwanted shadows.

■ It can cover or fill harsh shadows produced by direct lighting.

Negative effects of soft lighting include:

■ It flattens modeling of subjects when you want strong contours and detail.

■ It is difficult to control.

Lighting Contrast

Contrast, the relationship between the dark and light areas in a composition, is an extremely important aspect of quality lighting. A bright background can cause a subject to be recorded as black or extremely dark, obscuring facial expression or other details. A person standing against a brilliantly lighted white wall may appear too dark to see any visible detail. Areas that are too bright can also wash out color and cause other very unexpected results. Poor contrast can be avoided by correcting over-bright lighting or changing the camera placement.

Light Color and Color Temperature

Another key element of lighting is the color and color temperature. You will notice that light from different sources is registered as different colors. The basic color spectrum of a light source is most often measured by degrees Kelvin. This measurement has nothing to do with the actual temperature of the lamp but rather the color that a standard laboratory lighting instrument emits as it is heated to various temperatures. For example, both a candle and an airplane searchlight are direct light sources, but candlelight has much more orange and yellow in it. It appears warmer and, in fact, has a lower Kelvin rating. The searchlight appears whiter, brighter, and has a correspondingly higher Kelvin rating.

Light Source	Degrees Kelvin
Standard candle	1930K
Household incandescent lamps (25–250W)	2600–2900K
Studio tungsten lamp (500W–1000W)	3000K
Studio tungsten lamp (2000W)	3275K
Quartz or tungsten-halogen lamps	3400K

TABLE 4-1   Typical Color Temperatures (in Kelvins)

Light Source	Degrees Kelvin
Photoflood	3400K
Fluorescent lamps (standard)	3000–6500K
Sunrise and sunset	2000–3000K
1 hour after sunrise	3500K
Sunless daylight	4500–4800K
Midday sun	5000–5400K
Overcast sky	6800–7500K
Hazy sky	8000–9000K
Clear blue north sky	10,000–20,000K

TABLE 4-1   Typical Color Temperatures (in Kelvins) *(continued)*

Even the same light source can have different color output depending on various circumstances. For example, sunlight is warmer near sunrise and sunset than it is at high noon, and therefore it changes color. Serious videographers have been known to tear their hair out if a shot goes on too long and the light changes color before they are finished with a subject. (Remember, when I say a light is cool or warm, I am not referring to its temperature on a thermometer, but to the color of the light as it appears to the eye.) You can create the same kind of changes in artificial light sources. The standard daylight Kelvin rating for photographic or video lamps is 3200 degrees. This is indicated on the lamp package. This color temperature will match daylight to a reasonable degree. If you are using fluorescent lighting or standard warm lamps, you can compensate for the color temperature by using a filter over the lamps or on the front of your camcorder lens. I will consider color and the video camcorder later in this chapter.

Lighting Color as an Effect

Almost everything I have said about lighting up to this point is in the context of brightness—that is, the black and white picture. The addition of color has always been tougher for traditional video, but for desktop video it is actually easier in many ways. To use color effectively in video, you will find it helpful to understand some basic color theory.

Hue   *Hue* is what is commonly thought of as an actual color: red, blue, green, yellow, and so on. Red has been the traditional bane of analog video production. It is the hardest color to reproduce and control. For a variety of technical reasons, video red tends to "bleed" or "bloom," which is to say, spread beyond its borders. Not only is the rose red, but also some of the space around the rose. For this reason, reds are carefully used or often avoided altogether in video production.

Within its environment, digital video can handle most hues without such problems. Still, you need to be careful when recording your master footage and when your final output is on tape. Remember that your computer may be able to handle all hues, but not your TV set. Some editing programs allow you to select and output TV-friendly colors.

Saturation   Saturation is a measure of quantity, how much color is present in a given hue and how far removed it is from a neutral gray. This is what makes the difference between pink and red. Bright red is heavily saturated. Light pink is much less saturated. Heavily saturated colors tend to bloom in analog video, just as reds do. Digital video generally handles saturated colors well, but again consideration must be given to the final output format. Is your finished product coming back out to tape? Then be careful with saturated colors, because you want them to look as good on the TV screen as they do on your monitor.

Color Compensation   The fact that light can vary widely in color can cause continuity problems between your various scenes when editing a project together. When you have mixed daylight and artificial light, try to match the artificial light's color temperature with the daylight by using filters on your lens, or light-blue (also called dichroic) filters over your quartz lamps. If you balance your lighting to the incandescent lamps, the lamp will look normal and the daylight will look blue. If you balance to the daylight, the incandescent light will look orange. This can be an intentional effect if used appropriately.

Lighting Equipment

Light sources can be distinguished by the type of light they create (direct or indirect) and the quality and color of their light as well. Examples of direct lighting equipment include open-faced lights with exposed bulbs and fresnels. A *fresnel* is a spotlight with a spherical reflector and a stepped lens that produces a well-defined beam of light. Indirect lighting sources include scoops (inexpensive reflectors with or without clamps), softboxes (self-contained lighting devices that house light, reflector, and diffusion material all in a single unit), and direct light that is bounced off or diffused through something else. Warm lighting includes low-angle sunlight as well as tungsten bulb fixtures, which are the most common types of lighting instruments. Cool, whiter lights include direct overhead sunlight, as well as fluorescent and quartz tungsten-halogen lights.

For video lighting, it is best to purchase lighting designed and manufactured specifically for that purpose. In the long run, however, light is light and any safe light source will work for video lighting. Low-cost quartz workshop lighting fixtures, often provided with clamps, tripods, or stands, are excellent substitutes for more expensive lighting equipment. The color temperature of these lamps is acceptable for video, and professional lamps can usually be substituted for critical applications. Use what works!

Handheld or camcorder-mounted quartz video lamps, often called "sun guns" (see Figure 4-1), are excellent when used for interviews or for key or fill lighting on close-up shots. *Key lighting* refers to the lamp providing the principal source of light. *Fill lighting*, as its name implies, fills in what the key light doesn't cover. These lamps are usually powered by rechargeable batteries and are an invaluable and essential element of the complete light kit. Handheld lamps usually require a second person to hold them. As these lamps can also be cable-powered, they can have a much larger wattage. A disadvantage of camera-mounted lighting is the additional weight added to the camcorder. The camera-mounted versions, with battery power, are much lower wattage instruments.

FIGURE 4-1   Sun gun

Handheld lamps can produce a flat, washed-out image if the subject is looking directly into the camera. They work best when you shoot from an offset angle. Pay attention to the way the lighting looks in your viewfinder.

If you don't have a lamp with the proper color temperature, you can use a filter designed to compensate. It is called a dichroic filter and is clipped over the lamp to raise the color temperature when using the lamp in daylight. Dichroic filters can be found at video or photographic stores.

The open-bulb spot is a common fixture and is very efficient. These spots can be purchased singly or in kits and are relatively inexpensive. The professional fixtures (see Figure 4-2) come with a safety screen and can be fitted with flaps or barn doors and filter holders. Barn doors or flaps fit on lighting fixtures, and the amount of light can be controlled by the degree of opening or closing the flaps. The most commonly used lamps are 250–1000 watt quartz. Reasonable substitutions can be made for much less money by using workshop fixtures that are similar in construction. In this case, barn doors and other accessories must be homemade and added if required.

Scoops (see Figure 4-3) are common and useful, but not as efficient as open-bulb spots. They may be used to light backgrounds and used for fill. Scoops can be purchased at hardware and photographic shops. The photographic variety usually come with stands and stand holders or clamps. Every lighting kit should have at least two of these. They usually work with regular or photographic bulbs, reflector, or PAR lamps in wattages from 75–500. Photographic fiberglass diffusing fabric can be purchased at the photo shop and attached to the front of the scoop to create direct diffuse lighting.

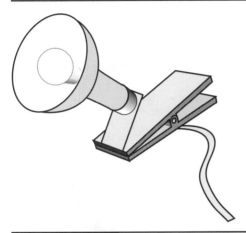

FIGURE 4-2   An open-bulb lighting fixture

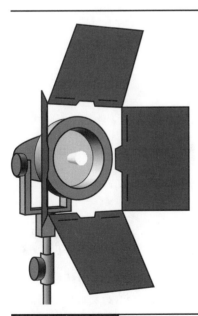

FIGURE 4-3   A scoop lighting fixture

Selecting the Proper Wattage for Your Instruments

Portable lighting equipment ranges from 75 to 1000 watts, and studio lights range from 1000 to 10,000 watts. A good all-purpose starter kit for the videographer on the road might contain three 500- or 1000-watt lights, with stands and assorted grip equipment. It's crucial to know the load your electrical circuits can handle. If you are limited to 15-amp household breakers, do not attempt to put more than one 1000-watt light into any single circuit. Although this may mean running extension cords, it is better than tripping breakers every few minutes.

The relationship between amperage (measured in amps) and wattage (measured in watts) is expressed in the following formulas:

Amps × volts = watts *or* watts ÷ volts = amps

Thus, if you have a 1000-watt lamp at 120 volts, it will draw 8+ amps.

For those of you who don't like math, here is a table of common wattages and their power consumption at 120 volts.

Lamp Wattage	At 120V
50W	.42A
100W	.83A
200W	1.67A
300W	2.5A
500W	4.2A
800W	6.66A
1000W	8.33A
1250W	10.4A
2000W	16.7A

TIP

An excellent choice for a basic lighting kit might include the following:

1 fresnel 1K

1 200-watt back light

1 500-watt soft light

These three instruments can safely be put on one 20-amp circuit (most common) and provide good flexibility when a diffuser is added for fill light.

Don't Short Circuit Electrical Safety

It's hard to pick up your Academy Award if you don't survive the production process. Safety should always be your first priority. Use your common sense and don't take chances. There are a couple of areas where the amateur, as well as aspiring professional video producer, should pay particular attention:

Equipment condition   Be sure your instruments, mountings, lamps, and cables are in good condition. Electric shock can occur from frayed cables or loose sockets and wiring. Broken clamps and fittings can result in burns or other injuries caused by falling instruments.

Grounding   Make sure that all your equipment is properly grounded and that three conductor plugs are not circumvented.

Lamps can become extremely hot and can cause burns. They can also explode under some circumstances. For example, rainfall reaching a hot lamp can cause it to explode. Always wait for lamps to cool before handling. Place safety screens over open operating lamps to prevent injuries should the lamps explode. Beware of fires caused by flammable material coming into contact with hot lamps.

Never touch the lamp (or bulb) with your fingers. The oil that hands leave on the lamp can heat and cause the lamp to explode. Always use gloves or a cloth to hold the lamp.

Besides using wiring that is safe and unfrayed, be certain to hide cables from traffic patterns or tape them down with duct tape or gaffers' tape to prevent falls.

Always keep a roll of gaffers' or duct tape in your video camera or lighting case. It is not only useful but will demonstrate to everyone how professional you are! Gaffers' tape is available at video supply or photo supply stores and is much preferred over duct tape because it is stronger, leaves no residue on walls and equipment, and is easier to work with. It also comes in a variety of widths, colors, and strengths that make it indispensable.

Putting Lighting Techniques to Work

Now you know what all that lighting terminology refers to, and what tools of the trade you need in your kit to produce good-looking video. You are ready to consider the techniques needed to put that knowledge to work!

Shooting with Available Light

Available light is just that—the light that is available in the area where you are shooting video. It may be sunlight or various types of indoor lighting. Available light is commonly used in producing video. Don't hesitate to shoot this way, but do be aware of color temperature in scenes shot at different times of the day. In addition, different lighting conditions or different types of lamps (such as fluorescent versus incandescent) can cause noticeable differences in color when you edit your video. Other things to watch out for:

- Extra care must be taken to make sure subjects are not in shadow or not adequately modeled.

- Lighting level may be too low (such as at night) or too high (such as in the noon sun).

- Lighting will change as the sun moves in and out of cloud cover.

You can easily enhance daylight using handheld reflectors of white cardboard or of aluminum foil-covered cardboard to fill or highlight subjects. Look carefully at the subjects, background, and so on, and place the camera and subjects to take the best advantage of the available light.

If your camcorder has a white balance feature, you can set it to automatic adjustment and it will compensate for some of the light color differences. Check your operating manual for information on white balance controls.

Shooting in Daylight

Daylight is free and generally available as a light source. In many cases you will be shooting outdoors where daylight is to be reckoned with as the primary or only light source. This can be difficult for a number of reasons. You can't control the direction of the sun, so you must control the location of the subject. There are also times when you cannot control the subject either. Just try moving the Leaning Tower of Pisa! There are additional problems of clouds and time of day. Lighting as it relates to the subject is an endless and complex problem. The usual solution is to change the position of your camcorder. This can determine the amount and effect of the light on your subject.

Daylight quantity or intensity is also a major issue. There are generally four things you can do:

- Place the subject in the light.

- Wait for the sun to change position.

- Move around your subject until the sun is behind the camcorder.

- Use reflected light to fill.

Using reflectors is a great way to compensate for uneven day lighting situations. Reflectors can be anything from a white painted wall to a professional video lighting reflector from the lighting supply store. The most common thing used by both amateur and professional videographers is white foam-core board from the art supply store. (Foam-core board comes in very large sheets, and consists of a thin lightweight Styrofoam core covered on both sides with smooth white cardstock paper.) It can be used as it is for a soft reflected light or you can glue aluminum foil on one side for sharper shadows and stronger light. It doesn't matter if the foil is crinkled and not perfect. Bringing one or two of these reflectors to your outdoor shoot is a good idea if you are shooting people or smaller subjects.

4

Reflectors are also great on indoor shoots when you need a little more intensity on a subject and don't have an extra light, or if the light source is too intense and you need to diffuse it. You can purposely crinkle and uncrinkle the foil before placing it on the board to provide softer reflected light.

Using the Direction of the Light Source for Effect

The direction of the light source, and where it is in relationship to camera placement, can have a significant effect on video quality. Consider the following four types of effects created by placement of the light source:

- Backlighting (lighting from behind the subject) can cause excess contrast and result in a dark or black subject.

- Straight-on front lighting can wash out facial modeling and expression.

- Lighting from below the face can cause the subject to look strange and bizarre. (Think of kids on Halloween holding flashlights under their chins.)

- Overhead lighting can also cause unusual shadows.

Move the camera or the primary source light (key light) until you are satisfied with the results.

Three-Point Lighting

Three-point lighting is the most basic lighting setup. Three-point lighting does not guarantee good lighting, nor is it necessarily boring lighting, but it is a point of departure. This section will introduce some terms and concepts that can be applied to any setup. When in doubt, go with the tried and true three-point setup, and experiment from there.

The three-point setup is a model for lighting an object or person in a way that provides definition and perspective (see Figure 4-4). As the name suggests, this type of lighting setup involves three separate light sources. There are no dark spots; light is provided from all angles.

Because the light is not even, there are differences in the lighted surfaces, which make the picture interesting and create a sense of depth and modeling. The following are generalizations from which any number of perfectly acceptable variations can be made.

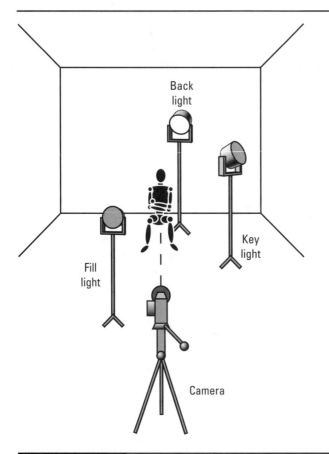

Three-point lighting setup

The first and most important light in the three-point lighting system is the *key light*, the major source of light on the subject. The direct line between the subject and the camera is called the *subject to camera axis*. The key light is a direct light source, placed within the 45 degrees of the subject to camera axis and slightly higher than the camera.

The fill light is placed on the opposite side of the camera from the key light to fill in what the key light doesn't cover. Usually it is a diffused light source with 40 to 60 percent intensity of the key light.

The backlight separates the subject from the background. It is a direct light source placed high and behind the subject. Make sure this light is tightly focused so that stray light does not reach the camera, causing lens flares. Other names for the backlight are *separation light*, *rim light*, *hair light*, and *kicker*.

The Magic of Bounce Lighting

Bounce lighting refers to the use of diffused sources as the primary means of lighting a subject. The purpose of this type of lighting is to create a setting of soft shadows, with an even fall-off or gradual reduction of lighting level on the subjects or scene. It can be effective for highly reflective items like watches and cars, and is particularly well suited for digital video. The illuminated scene appears smoother, less blocky, and without drastic contrasts between light and dark areas.

Generally, bounce lighting produces a lower light level than direct light and therefore requires cameras that are sensitive enough to produce good results under those conditions. It is appropriate when you want an evenly lit scene with multiple subjects without emphasizing one over another. Sometimes you may want to emphasize one of several subjects, one person of a group, for example, or one object among several others. In this case, add a direct source focused on the primary subject.

Key lights are frequently used with bounce lighting techniques. The key light may be bounced into a reflective umbrella, off a large white board (foam-core works especially well), or even a wall. The key may also be expanded to a wide beam and directed through a large piece of diffusion material such as fiberglass fabric from a photo shop (technically a direct source). Diffusion material is available in different styles and thickness, with corresponding differences in how much light is filtered and in what pattern it is spread. To work effectively, this technique requires a large distance between the light and the diffusion material.

There are special self-contained lighting devices called softboxes that house light, reflector, and diffusion material all in a single unit. The idea is to make the light source as large and uniform as possible. The bigger the reflective source and the closer that source is to the subject, the softer the light will be. As we use reflectors in shooting with outdoor light, indoor lighting can be diffused using similar techniques. Under these conditions, the fill light may be nothing more than white foam-core catching some of the spill from the key light and bouncing it back onto the subject. This method will work only if the fill board is fairly close to the subject and is best suited for close-ups with little movement.

By using a combination of large bounce lights, it is possible to light an entire set with a minimum of harsh shadows and total freedom of movement for actors and camera alike.

When lighting a scene, remember that there is a minimum amount of light required to properly expose your shot. This is called the *base light*. The amount of base light required depends to a degree upon the camera and lens you are using. Less expensive cameras tend to be less sensitive in low light conditions or simply produce a very grainy picture in low light.

Lamp location is best done by careful observation, but there are a few rules of thumb:

- The farther away from a subject the lamp is placed, the lower the light level falling on the subject.

- If you move a lamp too close to the subject, you will reduce its coverage on the rest of the scene.

- The farther the lamp is from a subject, the higher it needs to be placed to create natural shadows and modeling.

- Don't place a lamp too close to the camcorder lens as this will only flatten surface modeling.

- Don't try to eliminate all shadows behind your subject, but rather strive for natural or intentionally dramatic shadows.

Summary

Lighting is an important part of good video technique and will have a major impact on the quality and feeling of the finished video. A variety of equipment and techniques are available to give you good control of lighting, whether you are shooting inside or outside. The intensity of the light can be changed by reducing the aperture of your lens or by moving the subject into shadow. There are also easy methods to reflect light to increase the power and coverage of available light, and to diffuse lighting to achieve softer shadows and more even coverage. If all else fails and it is possible, come back at another time or day when the light is better. This introduction to lighting for video provides only a taste of the subject. Because lighting is one of the essential skills necessary to creating quality video for any purpose, you will want to experiment as well as consult some additional sources if your video projects warrant it.

Chapter 5

Record
Quality Sound

How to...

- Understand the basic principles of sound and sound recording
- Record sound for your video production
- Select music and sound effects for your videos
- Select the appropriate sound equipment for your video project

Video is really about active information or information in motion. Sound is *the* element that gives a fourth dimension to information presentation. It is also the easiest element to create and add to any presentation or project. Sound files recorded by the user are no more difficult to integrate than still photographs or blocks of copied text. Creating sound files is within the scope of anyone who has used a computer and tape recorder.

A video is made up of both images and sound, and only when these two things are in harmony have you succeeded in creating a good film. Lousy sound can ruin a video just as completely as poor photographic quality. Viewers may be directly conscious of images, but they absorb sound subconsciously—and it is precisely this subconscious perception that determines whether a viewer likes or dislikes a video. Sometimes you can get so wrapped up in the images you are taping that you completely forget there is also sound. Even professional videographers sometimes get annoyed when the sound person orders everyone to keep quiet in order to record the ambient sound of the particular location. While this is, of course, relevant to very advanced projects, sound can be a terrific enhancement to simple projects as well. We've all seen old home movies of kids playing or having a birthday party—silent except for the sound of the projector. Those days are gone! Now you can hear "Happy Birthday" sung as well as the sound of children's laughter. The improvement is tremendous, even for a simple video project as this. Today's easy technology allows even amateurs to leave the "home movie" mind-set behind, and develop good quality, entertaining, and well-edited video projects. This chapter addresses both the principles of sound recording and its application to video.

The Nature of Sound

Sound is the most basic means of human communication. It is the information bottom line. Psychologists assert that deafness is a much more profound disability than blindness. This is not to argue against the importance of visual images, but to recognize the fundamental nature of auditory perception and the important part it plays in human communication.

In video and film, you have a variety of sound possibilities—dialogue, special sound effects, ambient sound (the background sound at the event or location in which you are videotaping) or *Foley* (the addition of sound effects reinforcing the natural sounds that are made as part of the action, such as footsteps, door creaks, and so on), and, of course, music. Sound is just as important as the video portion of the content. In many cases, the soundtrack is much more important. In news, training, and documentary productions, for example, the audio track carries the majority of the information being imparted, with the visual components functioning primarily to illustrate the words.

Properties of Sound

A thorough knowledge of the properties of sound will form a basis for understanding how to record, edit, and process the sound in digital form. This is because digital processing and editing directly manipulate the basic elements of the physical sound, amplitude, and frequency. Consider that sound is made up of many parts: sound waves, amplitude, frequency, and timbre. Each is examined here.

Sound is physically created and transmitted as a wave within a medium (see Figure 5-1). The primary medium of useable sound is air, with secondary media being more solid objects in the environment. A vibration within a particular range of frequencies is set in motion by any number of causes, such as human vocal cords and musical instruments. These vibrations are transmitted to the environment through a complex game of "tag" wherein the sound radiates away from the source and agitates one molecule after the next of whatever it encounters until the energy is absorbed and the molecules finally come to rest. The sound creates waves that have a variety of strengths or *amplitudes* and vibrate at various rates or *frequencies*. All this vibrating and dancing of molecules through the air or other media is received by our ears as sound.

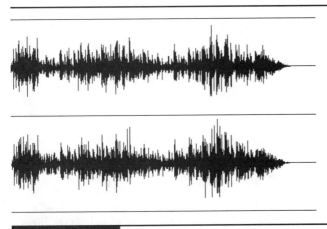

FIGURE 5-1   Illustration of sound wave forms

Amplitude

The amount of power that is present in a wave form, or in common terms, its volume level or loudness, is expressed as amplitude. In a practical sense, the amplitude is important at a number of levels in video production. The original amplitude of the sound must be loud enough to be distinguished by a microphone from the background or ambient noise. The amplitude must also be within an acceptable range for the limits of the transducer or microphone that is converting it to an electrical signal that can be recorded by the video or audio recorder. The electrical signal produced by the microphone must also be within the proper limitations of the recording and processing equipment. Care must be taken in the proper selection and placement of microphones and the use of either manual or automatic volume control that processes the electrical signal created by the microphones. The human ear can perceive and compensate for an extremely wide range of sound levels. Electronic equipment can't compensate as easily and readily for as diverse a range as the ear. It is important that you pay close attention to both low and high sound levels during recording and editing in order to achieve proper signal-to-noise levels on the low side and prevent distortion on the high side. The *signal-to-noise level* is the relationship between the actual sound you want people to hear and pay attention to, and the background sounds and electronic noise inherent in all electronic circuits. If you record the audio signal "hot" (as loud as your camcorder will record it without creating distortion), you will have the optimal signal-to-noise ratio. The human ear will concentrate on the information and ignore the noise or background as if it wasn't there.

Here is a chart that gives several examples of typical levels for familiar sounds. The intensity of sound level is expressed in volume units or decibels (db). This is displayed on audio equipment panels as VU or volume unit meters. The higher decibel values are the louder sounds.

Sound	Decibels
Jet aircraft taking off	125 db
Niagara Falls	95 db
Heavy traffic	85 db
Normal conversation	60 db
Quiet whisper	20 db
Outdoor silence	10 db

Frequency

The pitch, also called the *frequency*, means how high or low the sound is (see Figure 5-2). Frequency is measured in *hertz* (Hz). The audible frequency of sound is 20 to 20,000 hertz. This is the range of human hearing. Men with normal hearing can hear 16 to 16,000 hertz, and women can generally hear higher pitches as well.

20 hertz is a very low frequency sound, such as the lowest notes of a pipe organ or the rumble of heavy machinery. In musical terms these sounds are *bass* tones. High-pitched sounds include the chirping of birds or the highest notes on a violin. In musical terms these are *treble* tones.

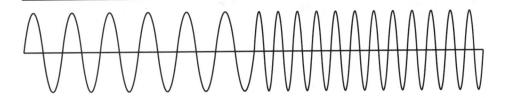

FIGURE 5-2   Frequency at 60 Hz and at 120 Hz

Timbre

Sound wave forms rarely consist of a single, pure frequency. They usually consist of several primary frequencies and multiple overtones. This property of sound is referred to as *timbre*. Timbre is the characteristic of sound that is its "personality." The timbre of a violin is significantly different from the human voice even when they are producing a tone of the same pitch or frequency, and the same volume or amplitude. This characteristic allows us to identify sounds.

The timbre of a wave form consists of primary tones and secondary tones called overtones or *harmonics*. A harmonic is a tone that is a multiple of the primary tone (2x, 4x, and so on). A primary tone can have one or many harmonics. The presence of particular harmonics may not be immediately evident to the listener, but the perceived quality of the tone is affected by them. The presence and quality of harmonics in a wave form is affected by the frequency response of the microphone recording it and the electronic system storing or transmitting it.

Amplitude/Frequency Relationship

Humans do not perceive all frequencies of sound equally. The ear is more highly attuned to the frequencies of speech. As a consequence, sound that is both higher and lower than these optimal frequencies needs a boost, electronically, to compensate. Similarly, microphones don't translate all frequencies with exact fidelity. This is true of every stage of sound processing. Both human and electronic idiosyncrasies are compensated for in a variety of ways. There are standard electronic circuits that balance recorded music for the normal human ear. An example of this kind of control is the Loudness control on an amplifier. This boosts the bass response as the volume is reduced to fool the ear into hearing the same volume relationships as at higher volume. There are also several electronic tools, both analog and digital, that allow the user to customize the control of the signal. They range from the simple bass and treble controls on amplifiers and mixers to sophisticated software algorithms that process digital wave forms in an infinite variety of ways.

Three Ways to Add Sound to Your Videos

There are three primary elements of sound in video production: dialogue (including voiceovers and lip synching), sound effects, and music. These elements are sometimes fully automated and sometimes mixed with live elements. An example of this is the use of recorded or computer-generated music to enhance a live lecture or presentation.

Dialogue, Voiceover, and Lip Synching

There are several types of voice sound recording. Dialogue, of course, refers to the words spoken by actors or presenters while they are visible on video. Although dialogue originally meant conversation between two or more people (one person talking was a monologue), today it is generally used to describe one or more people talking in a video or film. Dialogue can be your children telling Santa what they want for Christmas, or your parents reminiscing about their growing-up years and the way the world has changed. Dialogue can be your boss discussing the future of the industry, or a sales manager describing the virtues of a product or the direction marketing is taking.

Voiceover is narration, where the voice of the narrator is heard while the viewer looks at pictures or text. This can be an effective technique where actual dialogue is impossible—say, in the case of video of a vacation at a famous or particularly interesting spot. One friend of mine has a very effective video of a visit to Mesa Verde in Colorado. She videotaped these ruined cliff dwellings as she and her group hiked and climbed over some pretty rough terrain, gasping for breath—not ideal for recording good narration. Later she added a voiceover track describing some of the history and mystery of these dwellings. The finished video is a considerable step above the usual "what I did on my summer vacation" type of video.

Lip synching is the technique where actors or singers move their lips as if singing or speaking, but the actual sound of their voices is prerecorded and played back, accompanying their "performance." One of the funniest home videos I ever watched was a lip-synched version of "Cry Me a River" by a group of dressed-up and uninhibited teenagers.

Another type of lip synching is automatic dialogue replacement (ADR). This is the technique where actors re-record the dialogue of a scene already filmed or taped. The actor must perfectly match the new dialogue with the lip movements already captured by the camera. This is done because it is generally impossible to record perfect sound during the actual shooting of a scene. While many people are not interested in creating video as a movie, and therefore will never need to do this type of dialogue replacement, the editing packages you will be using do allow for this type of work. And, who knows, you may love working with video so much, you may decide to investigate the world of film making with your camcorder.

Sound also reinforces attentiveness and retention of material. Someone watching a video about how to do something can gain only so much from the visual image. Voiceover or dialogue about the task holds the viewer's attention and reinforces understanding of the tasks involved.

Sound Effects

Sound effects add depth and texture to a visual experience, increasing the effect of realism in a video presentation. A snapped twig, a creaking door, and footsteps on a dark street—these sound effects create a sense of suspense or menace in a way that often exceeds the impact of visual images. In the days of radio, sound effects were the primary tools of creating atmosphere, and how well they worked! No one who listened to *Fibber McGee and Molly* will ever forget the "closet," the continuing sound gag of McGee forgetting and opening the door of the junk closet. The resulting cacophony of crashes and smashes as the unidentifiable avalanche of junk came cascading out of the closet was good for a laugh week in and week out. The sounds, plus the imagination of the viewer, were more than enough. Sound effects are an absolutely essential element in the sound designer's bag of tricks, but they can also be terrific enhancements to amateur video projects.

Music

Music is tremendously important in video production. You often learn a great deal about what you are seeing in a film or video from the type of music that precedes or accompanies the visual action. The best example of this occurs within the first ten seconds of a movie. From the first few bars of music playing at the beginning of a film, the viewer is able to predict the mood and have a pretty good idea about at least some of the content to come. Repetition of musical themes is also used to signal mood changes or predictable events. The heroine has a lovely melodic theme, the bad guy a menacing one. Hearing either, you know who to look for next, even before they are on screen. No one who saw Steven Spielberg's film *Jaws* will ever hear that repetitive phrase of music—*bum* bum *bum* bum *bum* bum—without looking around apprehensively for the shark. Every composer who scores film and television knows how to create emotional effects and enhance the effectiveness of the experience for the audience. There have been several experiments by psychologists and sociologists pointing to a direct relationship between the music and the emotional effect of the film on the audience.

My experience of music in communication and psychological research supports the idea that it enhances both aesthetic experience and overall effectiveness of communication. There can be no doubt that skilled communicators in both business and the arts can improve their presentations with the use of high quality music. Music engages the listener in a way that visual information cannot. It allows the listener to think and integrate the information directly in the rational thinking process, a cognitive process that is an analog to speech and sound versus images. In fact, words and sounds can create powerful images.

Gauge Your Room Acoustics

Often the acoustics of a room can negatively affect sound quality. The most common problem is that of sound reflections from hard surfaces. This can be particularly difficult if you are shooting video in a public space where there are lots of unrelieved hard surfaces. Churches, community centers, and public buildings all can present problems of sound reflections. The use of absorptive materials such as packing blankets can reduce this problem. Moving microphones closer to the subjects can also help.

Using a Microphone

The microphone is one of the most important video production tools. It is the primary way sound, dialogue, and music are recorded on the soundtrack. In case you are wondering how it could be done without a microphone, remember that MIDI or computer synthesizers don't require microphones. Virtually all camcorders, in the consumer area, come with built-in microphones; these are usually adequate but not ideal. You will be better served by using an external microphone for shooting video where good sound quality is desirable. Generally, a stereo mini-plug microphone and line level input are provided on camcorders for plugging in microphones or external sound sources, such as tape recorders or public address system line outputs. More sophisticated professional microphones with XLR or quarter-inch phone plug cables will require an adapter or intermediate mixer between them and the camcorder. Please check the instructions with your camcorder and microphone for more information on your specific situation.

 *Buy a second microphone and plug it into the audio input of the camera. Be sure to test a microphone before purchasing it since microphones can vary greatly in their recording characteristics.*

The use of microphones is an essential skill, and it is important to know how to select the best possible one for your use. There are a number of microphone types, varying in both the type of transducer (the manner in which they convert sound waves to electrical impulses) and the directional characteristics of the individual microphones. We'll discuss both aspects in the following sections.

Types of Microphone Transducers

The transducer within the microphone converts sound energy to electrical energy in a form analogous to the original sound wave. It is made up of mechanical, electrical, and electronic components. The two principal types of microphone transducers are the dynamic or moving coil, and the condenser.

Dynamic or moving coil microphone   The dynamic or moving coil microphone is the most widely used in video production and audio field production. It produces a high quality signal, particularly for voice and field recording, and is very rugged. The frequency response in a professional quality microphone is good and it has good sensitivity. It comes in a wide variety of directional formats and physical packages. One of its strengths (besides its rugged nature and sound quality) is that it is the most inexpensive of the professional microphone types. The dynamic microphone requires no power supply.

Condenser microphone   The condenser microphone shows up in many forms, from high quality studio microphones that are very expensive to extremely inexpensive microphones for amateur use (these inexpensive versions are often called electret rather than condenser for some reason). In general, professional condenser microphones are designated as omni-directional, because of their sensitivity pattern which records sounds equally well in all directions. They are less rugged in field usage than dynamic microphones. The frequency response and sensitivity makes them very important for musical recording in the studio as well as a variety of general purpose uses. Professional quality condenser microphones tend to be much more expensive than dynamic microphones. Condenser microphones require a power source, either internal or external batteries or power supply. In many cases, they can be powered by the phantom power (a feature of professional mixers that supplies power for microphones through audio cables) supplied by professional mixer boards and consoles directly over the microphone input cable.

Directional Quality of Microphones

Almost all microphone manufacturers indicate the pickup pattern of their microphones. The shape of this imaginary space—for example, a sphere—indicates a pickup pattern of a microphone that is comparable to the area of vision of our eyes.

Omnidirectional   The omnidirectional (or nondirectional) microphone picks up sound equally well from all directions in the 360-degree area around the microphone, in an ideal theoretical sense (see Figure 5-3). In practice, omnidirectional microphones pick up sound from most directions. The omnidirectional microphone is useful for recording live musical performances, groups of actors or singers, ambient sound, and similar applications. Video field recording situations seldom call for omnidirectional microphones.

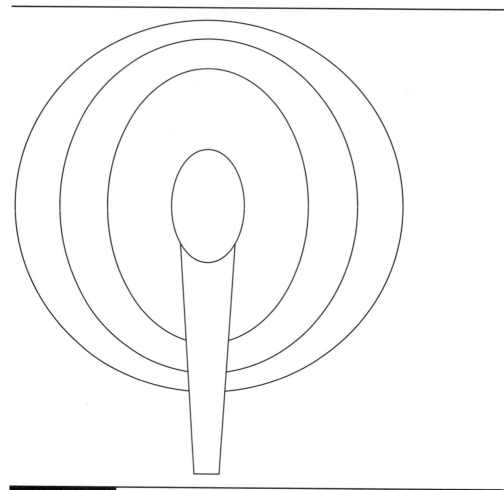

FIGURE 5-3   Pickup pattern of omnidirectional microphone

Unidirectional   The unidirectional microphone is sensitive in only one direction. There are a variety of pickup patterns that range from wide-angle cardioid to extremely narrow angle parabolic reflector microphones.

Cardioid   The cardioid microphone, by far the most common type, has a sensitivity pattern that is heart shaped (see Figure 5-4). A hyper-cardioid is commonly called a shotgun microphone. A cardioid microphone tends to record less ambient noise and is thus preferred for interviews, along with unidirectional microphones.

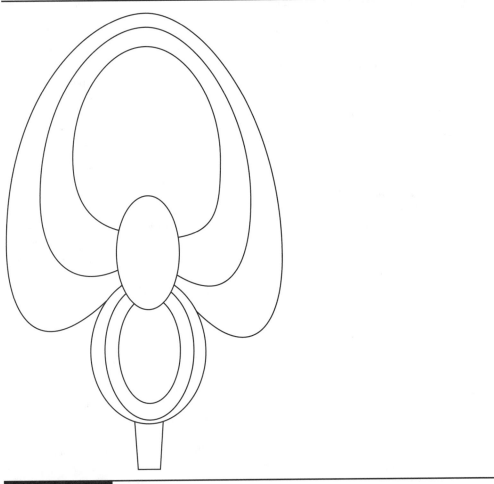

FIGURE 5-4   Pattern of cardioid microphone sensitivity

Microphone Application and Placement

The placement of microphones is important. A microphone placed too close to a source can pick up breath noise and distortion. A microphone placed too far from the source can pick up ambient or reflected sound that makes it hard to distinguish the intended focus of the recording from the background noise. The type of microphone selected can work for or against your intentions. The use of an omnidirectional microphone in an on-the-street interview can pick up too much noise and drown out the subject, where a unidirectional or cardioid microphone might eliminate this problem.

For recording ambient sound, use a microphone with omnidirectional recording characteristics. Even slight winds can cause an annoying whistle if you are using an unprotected microphone. To get around this problem, sew a cover out of a piece of fake fur and put it over the microphone. Even better, buy a wind sock, a foam cover for your microphone available in electronics stores. Since it barely cuts off the frequency response, you can conduct interviews even under windy conditions.

For interviews and people shots, a cardioid microphone is best as it is relatively insensitive to background noises. A lavaliere, or tie clip or lapel microphone, connected to an auxiliary recording device (MC or DAT) is also helpful.

Personal microphones are those that are attached to a subject. They are generally either clipped on to the subject's clothing (clip-on microphone), or suspended on a cord around the subject's neck (lavaliere microphone). These can be either wired or wireless and are primary tools for the videographer. Personal microphones are designed to be placed within about a foot of the subject's mouth. Pay attention to jewelry or clothing that might brush against the microphone or produce sound that might be picked up by the microphone. As always, listen to the signal to check the results of the placement.

Handheld microphones are used to record on-camera subjects, particularly in high noise situations. They are also good for news-style interviews. Be sure to select a cardioid pickup pattern. Hold the microphone at about a 30-degree angle and at a distance of 8 to 16 inches to eliminate sibilance or popping from the subject's breath. The use of windscreens over the microphone pickup area can reduce this problem, as well as noise from wind. Some microphones have proximity effects settings that may be used to eliminate this problem.

Pressure zone (PZ) microphones are omnidirectional and rely on reflected sound. They are placed on tabletops or stage floors, and serve as pickups for group interviews or musical recording. They can provide a high quality signal, but their omnidirectional pickup pattern means they are not selective in the sound they acquire.

Microphones mounted on headsets are useful for sports and off-camera narration. Be sure that the microphone is designed for video production and is equipped with windscreens.

Wireless microphones may have self-contained transmitters or may require the use of a standard microphone plugged into an independent transmitter. They can be used on any type of microphone desired. The receivers may be mounted on a camera. Their range is generally from 50 to 500 feet depending on conditions. Interference from commercial radio and television transmitters can be a problem. The selection of transmitter frequency and its use can be complex. Be sure to follow the manufacturer's instructions in using them.

Off-Camera Microphones

Up to this point we have been discussing the use of on-camera microphones. The use of off-camera microphones is of equal importance. They eliminate the problem of seeing the microphone in dramatic works and allow more unrestricted movement of the subjects. Their use can also reduce the number of microphones required when there are multiple subjects.

Microphone "fishpoles"—microphones attached to the end of metal or fiberglass rods and held by an operator—are an excellent solution to recording a single subject or interview. This method requires the operator to use headphones to ensure that the pickup and direction of the microphone are correct. This is also a standard news-gathering technique where proximity to the subject is difficult.

Recording Stereo Sound

People hear in stereo. The stereo soundtrack allows the simulation of space and realistic ambiance. It also allows more control of emphasis in complex soundscapes. Surround sound utilizes four or five tracks with speaker locations in front of and behind the listener/viewer. It is common on broadcast, laser, and DVD disc programming. Most of the stereo and surround effects are created in post-production (after the sound is recorded).

Stereo and surround sound are the standard ways to create soundtracks for commercial distribution. Microphones and microphone placement are important for ambient sound, musical recording, and subject location. When stereo or surround sound is critical to a production, it is usually created in post-production. If location or studio production requires special attention to stereo sound, it is best to simultaneously record it, with a time code track, to a multitrack recorder. When clips of small groups or individuals are being recorded, standard microphone placement is usually all that is needed. Most sound effects and Foley are also added in post-production.

For simple stereo recording, quality stereo microphones can be used. They give little separation, but stereo microphones are very useful for recording interviews. In the case of interviews, stereo separation is not essential.

Another option is to use two microphones arranged across each other at a 45-degree angle, forming an X. This is a good arrangement when stereo recording is necessary. It does not translate as well into mono.

A literally realistic sound perspective is impossible in video and film because of shifting camera angles and quick edits. The objective is to create an aesthetically pleasing one. One of the quickest ways to annoy your viewer is to include too much movement in the sound. It should be rich and complex, but not confusing.

Control of Dynamic or Sound Level

The most important thing about using external microphones is the control of the dynamic level or volume of the microphone signal being sent to the camcorder. A signal that is too strong for the camcorder will distort the soundtrack to the point of being unusable. Digital sound, as is universal in digital camcorders, is particularly unforgiving in this aspect. If you send the proper level signal from your microphone, mixer, or other external source to your camcorder, this generally will not be a problem.

All camcorders have automatic sound level or gain control that sets the audio levels so they will not be too soft or loud. Some more expensive camcorders also have manual audio level controls. This overrides the automatic control and allows the levels to be adjusted by the user. This gives greater flexibility and prevents unexpected level changes that can occur when sound levels vary greatly in shooting.

It is strongly recommended that you use headphones so that you can actually hear what your microphone is picking up. This will make it easy to make much more precise adjustments.

Sources for Sound Effects and Music

Professionals who want to add music and sound effects to a video generally have an in-house composer or sound engineer who can produce the required audio materials. Unfortunately, this is not practical for the amateur who wants to experiment with video production. There are alternatives!

License Sounds from Music and Effects Libraries

Sound effects libraries include collections developed especially for commercial distribution by major movie studios for their own internal use (in general or for specific movies). The movie studio collections are then selectively licensed for commercial use. Virtually all the major studios have licensed collections for distribution. The effects libraries may be general purpose collections with ambient sounds, crowd scenes, airplane and animal noises, and more. They may also be special applications, such as trains or military sounds, or collections for use in cartoons or dramatic productions.

The licensing restrictions for sound effects libraries are generally such that once the collection has been purchased, they may be used in an unlimited way in video or other productions with no additional fees to be paid.

Collect sounds and create your own sound archive. Every now and then you can also find a sound CD on sale.

Record Live or Custom Sounds

Often a video requires at least some custom or live sound effects recording. There are several rules for quality sound effects capture:

- Always use the highest quality microphones and recording equipment possible.

- Record as "hot" or at the highest signal level possible without creating distortions in the recording. This creates the best signal-to-noise ratio and ensures a clear sound with a minimum of background noise.

■ A realistic sound effect is not as simple as placing a microphone near a particular sound and recording. Microphone placement and room acoustics or wind can change the expected quality of the sound, and the end result may sound nothing like expected. Also, the actual sound of a particular object or event may not, when recorded, sound like the real thing at all. Many times the recording of something unrelated to the desired sound sounds more like the effect required. In other words, the real recorded roar of a lion may not sound nearly as impressive and lion-like as a composed sound effect made up of sounds from entirely different sources. In this case, art may surpass nature. Listen to the recorded result, and if the sound does not come across as expected, try again or try something else.

■ Sometimes an effect needs to be modified later in the computer to change the quality of a sound, or extend or contract its duration or pitch. Anything that works is correct.

Prerecorded or Stock Music

Stock music libraries, with music for every possible situation, are widely available with varying quality. The musical clips may be long compositions designed for background tracks for dramatic productions or documentaries, or they may be 10-, 15-, 30-, or 45-second clips for commercials. The biggest reason to use stock music is production cost. Stock music can be very inexpensive. The downsides include difficulty of timing clips to editing, overuse of the tracks in the marketplace, and difficulty finding just the right composition. Almost any style of music can be found, from a country jingle to classical music of the masters. These music library CDs can be affordably purchased for home use. Using music from commercial CDs is technically a violation of copyrights, particularly if the video is to be sold and distributed or played publicly or broadcast. Generally these situations don't come up in home video.

Synthesized Sound Effects

Musical Instrument Digital Interface, or MIDI, is a system that allows computers and electronic synthesizers to talk to each other and to record virtual performances of sound or musical compositions in a format that can be replicated without creating an actual sound recording. The term MIDI is used in a general, though not entirely accurate, sense to refer to synthesized music and sound effects generated by a computer or synthesizer device. The actual sound generation may be from a recorded collection of live or acoustic music, musical tones, or sound clips (called *samples*) stored in a computer or other device. Alternatively, it may be synthesized entirely electronically using a series of oscillators and sound processors.

Synthesized music and sound effects, particularly those controlled by MIDI software or sequencers, give maximum editing control. They can be minutely adjusted and linked to audio or video editing software for complete control. The major problem with MIDI-controlled samplers or a synthesizer is the inevitable synthetic quality of the sound. This can be an aesthetic limitation if the object is to achieve the sound of acoustic music. If contemporary sounds are required, there is no better solution. There are large libraries of MIDI sequences of custom, popular, traditional, ethnic, and classical music available as well as libraries of samples of every conceivable kind of musical instrument and sound effect. Thousands of MIDI sequences can be downloaded free from the Internet and used as-is or modified as required. If MIDI sequences of popular or current music are used, copyrights must be honored.

Summary

The multiple possibilities for good sound moves you far above the old silent home movie days. As you use your camcorder for a wide variety of projects, simple or advanced, you will want to explore the many ways you can use sound. Music, dialogue, and sound effects all can contribute to a finished, quality video project. Sound has a tremendous impact on the content of video and is effective if used appropriately and well. There is a wide variety of sound equipment, mixers, and microphones available in an equally wide price range for the imaginative amateur videographer.

Part II

Edit Your Footage and Add Special Effects

Chapter 6

Basics of Video Editing

How to...

- Plan your video production
- Plan your editing
- Create a script and shooting list
- Understand the editing process and its elements

In the introduction, you learned the importance of planning for the finished edited video. It may seem strange that we are spending the bulk of a chapter called "Basics of Video Editing" on the production planning process, but you will soon see the method to my madness. The primary point of video is to capture video shots that are used as visual "words" in the construction of a series of "sentences" or scenes that make up a "story" or finished program. The collection of words that you captured to videotape will be the entire vocabulary that you have to create your finished story. If you have forgotten any of the nouns and verbs, you will have a tough time completing the sentences, much less the full story. Often you will not have a chance to go back and capture additional video (your son's first birthday party only happens once!). Planning at its most basic level ensures the completeness of your visual video vocabulary, and at its best ensures that the quality of the finished video is high and no makeshift editing or visual Band-Aids are required.

Basic Editing Methods

There are a few concepts you will need to understand in order to successfully edit video. These include scenes, shots, clips, timelines, transitions, and project organization. There are two basic ways to approach the creation of a finished edited video:

The empirical or practical method   This method is based on the intuitive shooting of video using your instincts to capture scenes as opportunity presents itself. This instinctive approach is common to newsgathering and for the process of documenting unique real-time events. This is a valid method used frequently by professionals, requiring skill and experience. It is more difficult for the beginner to know what to shoot and what not to bother with, but it can pay off with big rewards if you learn to utilize it. You must be able not only to recognize important material and good shots as they happen, but also to anticipate them in order to be in the right place at the right time. You also need experience to anticipate what the video will look like and communicate in its finished form. A good way to try this method as a beginner is to take out the camera and shoot a birthday party, and then edit the scenes into a short video after the fact.

The planned method   This approach is based on a preplanned script or production plan that anticipates the objective of the video program and plans both the shooting and the audio track before the video is shot. This system ensures success and quality to a much higher degree, and is particularly recommended for the beginner. Filmmakers like George Lucas plan every second of film and every shot and camera move, almost on a frame-by-frame basis, before picking up a camera. Even though you are not making *Star Wars*, you will find this process of careful planning to be essential in creating polished, well-put-together video. This chapter focuses on the planned method. The things you will learn will serve you well even if you eventually choose to use the more empirical approach.

What Is Editing?

In the simplest definition, editing is the process of selecting and assembling a final audio or video production from the raw pieces. To avoid confusion I must point out that in the video world there are two meanings for editing. The simplest is the cleanup of videotape or film clips by selecting the beginning and end points and removing unwanted segments within the clips. The more global use of the term includes the first definition but encompasses all of the processes required to create a finished video or film project that tells a story or makes a point. The editor is the person who has the responsibility for the selection and final assembly of clips, titles, effects, and other visual and sound elements into a finished video or film. In its basic form, editing is storytelling.

In traditional editing of analog video, there are two primary categories describing the editing processes: offline and online. Offline editing is the electronic editing process during which an intermediate video is edited into a submaster and then an edit list is created for a final online edit session. Online editing is the final electronic editing stage where the final master is created. These are technical definitions, but the point is that in traditional professional editing there is a phase during which the individual elements are edited and then assembled into a final list of elements, which are then assembled into the final product, the master. There were technical and logistical reasons that this two-part process was developed. The high cost of using the full master or online edit machines meant that the editor used less costly edit machines for the time-consuming initial editing and reserved the more expensive machines for the final version.

In digital home video, the cost and time factors are not an issue. But the two-part process of organizing and editing the scenes first and then assembling them into a final product still makes sense from an organizational perspective. It is a logical and organized method and ensures that fewer mistakes will occur in the final master.

Why Edit?

You edit in order to make your point without boring your audience. It's just about that simple! If you shoot four hours of video during your vacation at the Grand Canyon, it is unlikely that your friends and family will sit still through the fourteen shots of a grizzly bear or of your daughter waving at the camera. You will do much better if you select the best two or three bear shots and only one of the daughter waving. After all, the video only represents the vacation, it can't have the same, or seemingly the same, duration. If you are using the video to teach your fifth grade class how to mow the lawn it is unlikely that a one-hour tape made with the video camera fixed to the tripod will either motivate or give enough information to be effective. It is important to break the process and the video up into the key steps with explanations of each. You might only need five minutes to explain how to start the mower, cut in a circle, and avoid the flowers. Add a few titles, some narration, and a touch of humor, and you will have the entire class working in the yard on Saturdays.

If you are going to make effective edited videos, you must, however, have good shots to work with. Using the "beginning to the end" (also called "in sequence or out of sequence") or linear shot method, you point the camera and turn it on, recording anything that comes in front of it. In many cases, this might be all you want a video camera for. It will record the baby's first birthday party, an anniversary, or a vacation at the beach. That is what most people who have a video camera use it for. These tapes usually go into a shoebox in the closet for later viewing. The problem comes when you want to share the video with others. For example, you may want to extract a clip to send to grandma over the Internet. You can use simple editing software for that, and planning isn't important.

Another way to share is to show the tapes to friends and relatives. This is where it becomes problematic. Videos like these may record the event, but they seldom convey to the audience any of the joy or interest of the actual events—not even to those who were there in person. The dull dead time you recorded between interesting shots seems to get longer and longer as the evening progresses. Editing can save the day, and it doesn't have to be complex. You just need to extract the good stuff and glue it together into a more interesting sequence. You might find that ten or fifteen minutes are used out of two or three hours of vacation shooting, or only three or four minutes from the first birthday party. But they can be a terrific three or four minutes!

Scene by Scene

If you want to tell a story or teach with video, you must shoot the videotape in an entirely different way from the linear shot method. It is certainly possible to extract scenes from linear shot video to use for teaching or storytelling, but the odds are you will come up short with shots required to make your point. The more successful way to create a teaching or storytelling video is to begin with the end. What I mean is that you need to decide what story you are going to tell or skill you are going to teach and design a series of shots and scenes that will accomplish your goals. In addition to reducing the chance of forgetting something important, shooting with a script is more time efficient. You can shoot all your close-ups when lights and camera are set for it, then move on to your medium shots and wide shots. Integral to preproduction is the creation of a *shot list*. During this process, the director groups shots of similar lighting/camera placement so they can be recorded at the same time rather than resetting over and over again.

Why Plan?

"Garbage in, garbage out" is familiar computer jargon that is also relevant to the world of video editing. If you don't have good material to begin with, the finished product is not likely to be good either. It is difficult to create a quality final video with randomly shot video clips.

The chance of capturing everything needed to tell a story in video is unlikely unless you think before shooting (or at least at the time of shooting) about what the finished product is to be. When writing a mystery novel the author usually knows "who done it" before setting pen to paper (or fingers to keyboard), and the creation of exposition and characters leads up to that final conclusion. A video is the same way—you have to know the outcome before you can create the scenes leading up to the end, and be sure you have it all when you get there. If you understand before you begin what you want your audience to know, do, or feel, the whole process becomes easier. Once this is established you can continually check back to make sure what you are doing maps to your program's objectives. An effective planning process for your video consists of three stages:

Preproduction   This is the advance process of planning a video, including a program concept, script, and other preparation.

Production   This is the process of creating and shooting the video elements for your program.

Postproduction   This includes creation of titles and effects, editing, and output to master tapes, CD-ROM, or Internet formats for viewing by your audience.

I am covering the preproduction and the postproduction process in this chapter. The production process is covered in Chapters 2 through 5.

The Nature of a Video

Take the time to watch different programs on television and pay strict attention to the elements and concepts introduced below. You might even use a pad and pencil to track and list the different elements as they occur. In a movie, the elements in the order they might typically occur are:

- Opening logos
- Titles
- Major credits
- Establishing or opening shots
- Dramatic scenes with characters

In most movies, there are several logos representing the distributor (such as Warner Bros.), the producer's company, the director's company, and more. Each of these logos, credits or titles was created as an individual element, a graphic title or animation. These are connected together to form a continuous bit of film.

Often after the opening logos and titles you will see a series of shots of a landscape or cityscape that establishes the location of the movie. This is called an establishing shot. Sometimes the titles are placed over the scene itself. This scene may feel like a single piece of photography, but if you watch closely as the camera moves or as the visual perspective changes, you will see it is really a series of individual shots that have been edited together to give the impression of a seamless continuous camera move. This feeling is created by successful editing. Editing is the process of assembling a series of different but interdependent video or film shots into a continuous linear presentation generally without bringing to attention the editing process itself.

There are certainly exceptions to these rules, but you will find that the exceptions are rare indeed. The same rules apply to every kind of video and film production. Two places where the edits themselves are deliberately brought to the attention of the viewer as a regular and deliberate element are the news and training films. Even here, good editing takes place within the individual clips. Only in legal videos designed for use in court and other legal contexts does the absolute demonstration of the clip as being unedited become important. In this case the videographer usually includes a running clock in the shot to prove that no editing was done.

The point of this book, however, is not to teach you how to make movies or shoot legal video, but to create good quality and effective video for personal use. The principles discussed above are equally effective in this pursuit.

Planning a Video

The best thing you can do is to begin with an idea. What is the objective of the video you are creating? Is it to show how much fun you had at the Grand Canyon? Once you have the concept, the next step is planning how you will achieve it. You need to think about what shots you need to capture to tell the complete story, what camera positions and angles will be the most interesting, what music will set the mood, and what words you need your subjects or actors to say or your narrator to record. All this is usually planned in some detail in a script.

The Script

A script represents the ideas and objectives of your video production. Take time to outline the flow of ideas you wish to present and think of both the visual context and words. Some scripts focus on the words and the visuals provide illustrations to the words; others rely heavily on the visuals with few or no words. Create a script format that serves your needs. Some points to remember:

- Video is a visual medium.
- Show images that communicate meaningful ideas and feelings.
- Don't bore your audience with static shots or shots of irrelevant things.
- Keep it simple and to the point.

- If you are using a presenter or narrator, try to establish his or her presence, but primarily shoot the subject or subjects. Use the narration over the more interesting images.

- If you are videotaping actors, remember that you should generally establish the context, then focus on the subjects, with shots of each important element communicating the action or dialogue.

- Look at and analyze similar video or film products for ideas on shooting and editing.

In film and television, there are formal script formats and standard terms and abbreviations. For fun, here is a list of the abbreviations used in making a script:

INT: Interior

EXT: Exterior

F/U: Fade up

CU: Closeup

WS: Waist shot

LS: Long shot

POV: Point of view

O/S: Over shoulder

LD: Lock down camera on tripod

You don't need to worry about these abbreviations for your script. Create a format that is useful to you, that contains all the relevant information, and that is easy for you to understand. If you decide to go professional, you can look up professional script-writing books in the library. I have included a simple script example below to give you an idea of how you might begin. First, you should give a brief overview of the objective of the script if it is a documentary or a synopsis of the story if it is a mini-movie. Next, make a list of the segments and elements for the video. This should be done in the order you are going to shoot rather than the order they will occur in the finished edited video. (This is important because it is often impossible to shoot in the order the scenes will appear.) Include appropriate information for each shot.

Visiting the Museum	Total Duration: 15 minutes	
Shot 4	Opening titles	00:15
Shot 5	Establishing shot of museum	00:10
Shot 6	Introduction in lobby	01:30
Shot 7	Exhibit One	07:00
Shot 8	Exhibit Two	05:00

Single Camera Script

FADE IN:

Scene 1.0 EXT: Front of Museum
Description: A slow zoom from a full shot of the building into the front entrance of the building. Show people entering.

Scene 2.0 INT: Lobby of Museum
Description: LD of the presenter in the lobby of the museum.

Presenter: Welcome to the Acme Museum. It was founded in 1948 to commemorate the 100th anniversary of Tinyville and contains a number of exhibits showing the life and work of its citizens during its early history. We will be visiting a number of the exhibits. Etc.

Scene 3.0 INT: First Exhibit
Description: Establishing shot of the exhibit. (An establishing shot is one that sets or establishes the context of the scene.) Pan from left to right.

Scene 3.1 INT: First Part of Exhibit
Description: First part of the exhibit. Pan across the exhibit. POV (point of view) museum visitor.

Scene 3.2 INT: Presenter
Description: Shot of presenter in front of exhibit.

Scene 3.3 INT: Second Part of Exhibit
Description: Pan across the exhibit. POV museum visitor.

Scene 4.0 INT: Second Exhibit
Description: Establishing shot of the exhibit. Pan from left to right.

Scene 4.1 INT: First Part of Exhibit
Description: First part of the exhibit. Pan across the exhibit.

Scene 4.2 INT: Presenter
Description: Shot of presenter in front of exhibit.

Scene 4.3 INT: Second Part of Exhibit
Description: Pan across the exhibit. POV museum visitor.

The first number in the sequence scene designator (such as 3.x) is the scene number and the second number of the scene designator is the shot number (such as 3.1). Include a description of the shot and dialogue, if any.

Storyboards

A storyboard is a series of drawings of scenes used to design and organize a shoot or edit. This can be a particularly useful way to organize a shoot as it focuses on the visual aspect of each shot. It can help plan the actual placement of the camera on a scene-by-scene and shot-by-shot basis. Storyboards also help plan the changes from one shot to the next to preserve picture continuity or the relationship of one shot to the next. A storyboard doesn't need to have fine art drawings. Simple pencil or felt-tip pen sketches are typical, and stick figures are fine (see Figure 6-1). You just need something to represent your ideas.

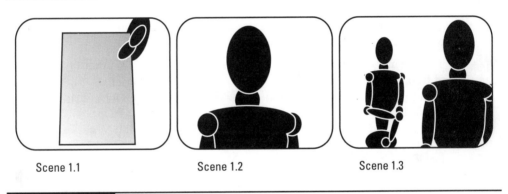

| Scene 1.1 | Scene 1.2 | Scene 1.3 |

FIGURE 6-1   Typical storyboard sketch

TIP   *You can find video storyboard sketch pads in your local art supply store. These pads have bubbles representing the TV screen and a place below each screen for dialogue and shot instructions. These are handy to use and will make you feel like a pro.*

The Editing Process

I know it seems like a long time since we began talking about editing, but as you will soon see, the majority of most productions are "edited" when you are shooting. The editing process is 80 percent selecting, trimming, and assembling the best shots into scenes and finally into the finished program. The remaining 20 percent is the addition of music, titles, and transitions that make the production sparkle as a finished product. That 20 percent is extremely important, but in most productions it takes less overall time. Let's recall the key elements of an edited video:

Scenes   The scene is the key "sentence" in your video story. It usually contains more than one shot. A scene in your vacation video might consist of the initial drive into the Grand Canyon park. It is more interesting to break this into several shots (perhaps a wide shot of the entrance, buying your ticket, a shot of the sign, driving down the entrance road). In a few seconds of video shorthand, you can show a 45-minute wait in line to get entrance tickets.

Shots   A shot is an individual "word" that you use to create a scene "sentence."

Clips   A clip is an individual shot that has been selected from your raw tape and captured into your editing computer, where it is stored and ready to assemble into a scene or program.

Timelines   A timeline is the software screen that allows you to assemble a series of clips into a video sequence and add such elements as transitions, titles, animation, and so on. The timeline is the primary editing worksheet. I will cover the editing process and the timeline in more depth in later chapters, beginning with Chapter 8.

Transitions   Transitions are the special effects that move from one shot or clip to the next, or from one scene to the next. The most commonly used transition—and the one you will want to use 90 percent of the time—is the cut edit. This edit simply connects the last frame of clip 1 to the first frame of clip 2 with no transition effect at all. One of the most commonly used transition effects is the cross fade, wherein scene 1 slowly fades away as scene 2 becomes visible. At some point you see elements of both scenes. Other transition effects include wipes, spins, zooms, and so on. You are so used to seeing these transitions on TV and in films you probably take them for granted. Pay attention the next time you watch TV and make notes of the different transitions in the program. Notice that the vast majority are simple cut edits. We will cover transitions in Chapter 8 and beyond.

NOTE   *To keep editing invisible to the viewer, most of the time you will want to use only cuts and dissolves. Nothing spells amateur like flashy effects for no reason.*

Titles   Titles are the segments of video that present textual information about the video. They might be separate animated segments or the text may be overlaid on live video. Titles are created with your editing software. You can also use other software including specialized titling programs, Adobe Photoshop, or other paint packages, and then import the titles into your video editing software. Titles are an important element in your video. I will cover the creation of titles in Chapter 11.

Effects   There are a wide variety of special effects used in video. They range from animated titles to morphing of one image or scene into another to blue screen compositing. In this last process, actors or other visual elements are shot in front of a special blue or green background that can be dropped out. Then they are inserted into another video screen with other images and the result looks like a single shot. An example of this effect is the TV weather person standing in front of an animated weather map. The person was actually shot in front of a blue screen, then later added to the scene with the weather map. They appear to be standing in front of the animated map as they gesture to incoming storms. Life can be, particularly in the age of special effects, an illusion! Titles and transitions can also come under the heading of special effects. In film and video effects are often called FX for short. I will cover effects in more detail in Chapters 11 and 12.

Soundtrack   The soundtrack is the other half of your video. Much of the video you shoot will have a soundtrack captured at the same time as the video. In amateur video documentaries and news gathering, this is important and you should capture the best quality of sound you can. You will need to pay attention to sound level, background noise, and intelligibility. Additional sound is often added in the form of voiceover, and, in most productions, you will add music at least in the introduction and titles and often as an accompanying soundtrack. I discussed sound production in Chapter 5, and I will cover sound editing beginning in Chapter 8 and in more depth in Chapter 10.

Editing on the Computer

In Chapter 7, you will be selecting the proper editing hardware and software for the computer, and then you will learn to use the computer and video editing software in Chapter 8. For now, let's look at the basic process of editing on the computer, which is generally done in the following steps:

1. Select the video clips from the raw videotapes using your shot list and script, as well as more detailed lists made from viewing the finished video. Note the beginning and ending times on your camcorder or VCR.

2. Capture the video segments into clips via the FireWire cable connecting the digital camcorder to the computer. The selection and installation of this equipment is covered in Chapter 7.

3. Trim the video clips so that the beginnings and ends are clean and appropriate. Make sure to leave "handles" on your clips so you can use a dissolve if you choose.

4. Create titles and other special animation or effect clips.

5. Import all video clips including titles and effects into the timeline.

6. Assemble and add transitions into the timeline.

7. Import audio tracks and elements into the computer.

8. Create music and narration tracks. Edit and trim an audio element in the video editor or external audio editing software or equipment.

9. Import audio elements into the video editing software and place them in the timeline.

10. Make final adjustments of time, duration, position, sound levels, and so on in the editing timeline.

11. Create a preview video to check the results of your edit. This is done many times during the editing process and can be done in short segments to save time.

12. Make last minute adjustments of the edit timeline.

13. Create final master video on the hard drive or tape for final use and distribution on tape, CD, Internet, or via a local area network.

I will cover each of these steps in detail in one or more later chapters.

Summary

This chapter covered the key concepts of planning your video project with editing in mind and planning your video editing process. The next part covers the details of the editing process from capturing video to creating finished video for tape or the Internet. Good planning will save you time from the start and ensure a much higher quality project in the end.

Chapter 7

Select Your Digital Video Capture Hardware and Software

How to...

■ Connect your camcorder to your computer

■ Select the appropriate editing software for your needs

■ Select a FireWire port

■ Install your FireWire port

Let's say you have taped the baby's first birthday and you can't wait to turn it into a movie and inflict copies on all the grandparents and relatives. Now, you realize that in order to edit video into an Academy Award winning production you must get the video you've shot into your computer and learn to use a video editing software package. This chapter will start you on the road to that objective.

Connecting Your Camera and Your Computer

The first step is determining what kind of connection you have between your camcorder and the computer. The second step is the selection and use of a video editing software package.

The connection between your camcorder and your computer is made in the digital video world using a FireWire or iLink connection. This is also called an IEEE 1394 port. These imposing names all describe the connector that allows the transfer of digital information from external devices into your computer—in this case, transferring video from a camcorder. Either your computer came with the necessary port or you'll have to purchase one and install it to do any transfer of video.

A FireWire Port by Any Other Name...

The Institute of Electronic and Electrical Engineers (IEEE, sometimes pronounced "I Triple E") creates many technical standards, including the standard that computer engineers use to design and implement the connection between your video camera and your computer. The document that describes the standard is IEEE 1394. The computer industry decided that it needed a sexier name for this port and dubbed it the FireWire port. Then Sony, in their usual independent style, decided *they* would call it the iLink port. So if you hear or read about a FireWire or iLink port—or an I Triple E 1394 port—it's all the same thing.

Whatever you call this port, your computer needs to have one to get your video from your camera to your computer. Many new computers come equipped with FireWire ports, and if this is the case you will only need the proper digital video editing package to begin. If your computer was not equipped with a FireWire port, you will need to purchase a third-party board and install it in your computer. When you go to CompUSA or other computer store, just ask for a *digital* video input card, which will include the FireWire port or ask for an add-on FireWire port card. Installing these pieces is not a very hard process, but if you're nervous about taking it on yourself, your local computer store (preferably the one you bought it from) should be able to install it for you.

If you are using a desktop computer as opposed to a laptop, the FireWire port will come on a PCI plug-in card and will be accompanied by the proper drivers you'll need for installation, as shown in Figure 7-1. The plug-in card and drivers you'll need are often bundled with at least a basic video editing package.

FIGURE 7-1   PCI FireWire card

If you are using a laptop computer that is not equipped with a built-in FireWire port, you will need to purchase a FireWire adapter that uses the PCMCIA card port or PC Card port (the same thing) in your computer. The PC Card (as shown in Figure 7-2) attaches to the side of your laptop. This adapter simply plugs into the PC slot, but will require the installation of drivers which will be included with the card you purchase.

The board you'll have to purchase ranges in cost from $100 to $1,000 depending on the bundled software and other features added to the card (such as analog video capture connections and video accelerators). For basic digital video (DV) editing work, the most basic FireWire port is all you need.

FIGURE 7-2	PCMCIA card

The entry-level products for both PCI and PCMCIA connection made by Dazzle are good examples of adequate ports and are available in most computer stores or via the Internet. Both products come bundled with basic drivers and an editing package.

More sophisticated software/card bundles include the DV.now and DV.now AV products from Dazzle (see Figure 7-3) and the DV-500 from Pinnacle systems. All of these boards come bundled with Adobe Premiere, a more professional and therefore expensive editing package. These higher end bundles range in price from $500 to $850 with the Premiere software. The DV.now AV product adds analog video capture to the FireWire connection. This is the product I used in writing this book.

FIGURE 7-3	DV.now AV FireWire capture card

Using Built-in Connections

If your computer came with a built-in FireWire or iLink connector, it won't require software or hardware installation—and this is a plus. You will still need a DV capture software utility to use remote control features of your camcorder (if your camcorder supports these features) and to transfer video clips to your hard drive for use in your edit package.

To connect your camcorder to the FireWire port, you will need a FireWire cable. The FireWire cable comes in sizes ranging from 18 inches to 6 feet. You should use the shortest cable convenient to you so you don't have miles of extra cable lying around and tripping you up. You should also be aware that there are two connector types on the end of FireWire cables: a small and a large connector. The large connector is usually present on computers and add-in cards as well as on desktop DV players. The small connector is usually found on camcorders. Be sure to check the size of the connectors on your computer and other equipment. Make note of the large/small connector type on your computer and device so that you purchase a cable that has the proper combination to match.

Connecting the cables is a simple matter. Just plug them into the respective sockets! When you turn the external device on, your computer will recognize that a DV FireWire device is available to programs and will usually notify you with an onscreen message. When you are using capture software, you will usually need to select the DV device you want to capture from with a menu option. There may also be an update command that checks the current DV devices plugged in and turned on.

TIP   *If your computer has problems recognizing your device, make sure the device is turned on. If the problem persists you may not have the necessary drivers properly installed. In this case, consult your camcorder manual or contact the manufacturer of the device.*

NOTE   *It will take 2GB of hard disk space for every 10 minutes of finished movie. Your hard disk drive must be capable of sustained throughput of 4 MB/sec. Fast hard drives capable of this kind of performance can be purchased as external drives with FireWire connections for as little as $250. They simply plug into the alternate FireWire port that comes with your FireWire card. Just ask the computer store salesperson for a drive to use with digital video editing.*

Adding a FireWire Port to Your Computer

I am using a Dazzle DV.now AV card on my Compaq Presario computer. The installation procedure was simple and similar to installing FireWire cards from other manufacturers. The steps are:

1.   Turn off your computer and disconnect the power cable.

2. Open the cover of your computer and expose the PCI slots. These are the connectors on your computer's motherboard with the cards, such as your audio or video connectors, plugged into them.

3. Remove the cover to the rear of your computer from one of the PCI slots. This is done by unscrewing the screw that holds it in place. Hang on to the screw—you'll use it to secure the new FireWire card in place. Be sure that the type and location of the slot conforms to the recommendations of your board manufacturer. Check the computer's user manual for instructions as to which PCI card slot to use for this kind of upgrade or refer to the installation instruction that came with your FireWire card.

4. Before removing your FireWire card from its anti-static package, touch the computer case to discharge any static electricity you may have accumulated on your body.

5. Remove the card from its package and plug it carefully but firmly into the selected PCI slot.

6. Replace the screw that holds the card in the slot.

7. Follow the manufacturer's instructions for installing and testing your drivers and other software utilities.

Anyone can accomplish this process without fear of disaster. Just follow the instructions carefully and keep a telephone at hand to call technical support if you get into trouble. You should always follow the installation instructions that accompany your FireWire card for specific instructions. I included the basics above to assure you that the process is within the capabilities of most computer users. For the extra cautious, before you do anything to your computer, always back it up.

After receiving my DV.now board, I discovered that the manual said the board worked with a Pentium III processor. Well, my Compaq has an AMD Athalon processor! I immediately called technical support to make sure I was not beginning a hopeless process. They asked me to open my computer and check the model number of the chip set. They then informed me that the board was compatible with my computer, but that older computers might not be compatible. Check with your dealer or the board manufacturer for processor and computer model compatibility, if possible, before purchasing your board. This will save time and tears.

Choosing the Best Editing Software for Your Needs

Once you have selected the FireWire connection, you need to select your editing software package. The software package that came with your card may be adequate, or you may need a different one to meet your particular needs or skill level. There are basically three levels of video editing packages that target the consumer market. Although they will vary in the type and number of features and special effects as well as their ease of use, your finished video will look exactly the same whichever you choose to buy.

Low-End Editing Software Products

The less expensive packages frequently come with clever interfaces, but perform only the most basic functions. Take a look at the software interface screen shown in Figure 7-4. It is presented in a very inviting way that resembles the controls of a VCR. This makes it easy to initially understand, but presents limitations as you get more involved in video editing.

This level of software is fine for the occasional user or for kids to learn on. For the more serious user, however, its limitations would be an obstacle. These packages usually come bundled with FireWire cards, but can be purchased separately in the $25 to $100 range.

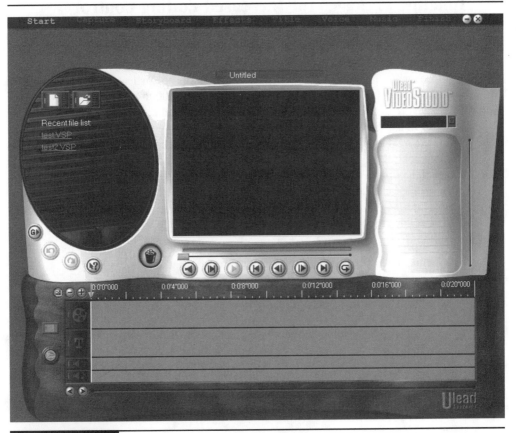

FIGURE 7-4   Edit screen of Ulead VideoStudio

Consider using a product such as Ulead's VideoStudio when you want to do the following:

- Convert any analog video source to TV-quality MPEG-1 digital files
- Edit video files with drag-and-drop ease to create custom videos
- Add special effects, transitions, scrolling titles, background music, and voiceover narration
- Create RealVideo or Windows Media streaming audio and video
- Record final videos on tape or burn CDs (CD recorder required)
- Create still image or video slide shows for school or business presentations
- Email video and still photos

Timeline Packages for Increased Editing Control

The second kind of video editing product available is a mid-level timeline-based package such as Dazzle MainActor and the limited or light editions of Razor Edit. These are junior versions of the more professional editing packages and use similar techniques. What you learn from these packages will allow you to easily upgrade to more sophisticated packages as you become hooked on video production. These packages cost from $100 to $250 and usually are purchased independently from a hardware/software combination bundle.

We will use the Dazzle MainActor package from this level in several chapters of this book. Notice the more complex screen layout in the software interface shown in Figure 7-5. It allows for more control of video, audio, and transitions such as fades and wipes. It is set up like more advanced and professional software editors, so the way you think about the project and the things you learn to do will easily transfer to more advanced software if you decide to move up to them later.

The following features are common to mid-level editing software:

- Special effects, titles, and transitions, including professional-quality features like video and audio filters, a built-in morphing module, and 2-D and 3-D titling.

- Professional-quality video codecs, including DV, MPEG, and Motion JPEG codecs for fast rendering. A codec is a software feature that processes the video and compresses it so that it takes less space on the hard drive or tape and allows more video to be recorded than would be otherwise practical. All video editing software includes some choices of codecs; the more expensive software packages include more and better choices.

■ Background rendering and smart rendering (DV and Motion JPEG formats), plus real-time previewing. This feature allows the user to process finished video for use while doing additional editing on the same computer, which saves time.

■ Ability to import and export AVI, MPEG-1, MPEG-2, QuickTime, and Motion JPEG.

■ Ability to export RealVideo and ASF streaming video.

■ Compatibility with a variety of video hardware, including low-priced IEEE 1394, FireWire, and iLink cards.

These features give more choices for playing your video on different computers and with additional programs.

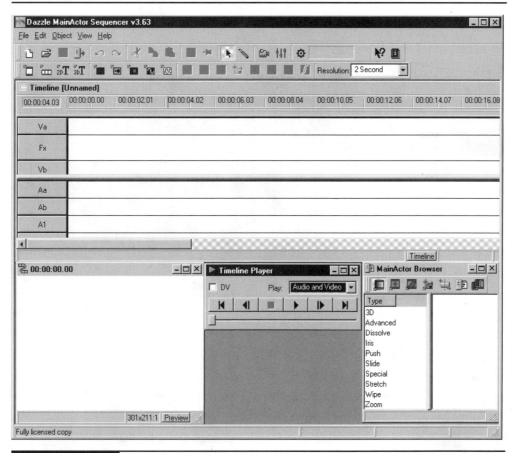

FIGURE 7-5   MainActor timeline editor

Professional Editing Packages

The third kind of video editing software package is called "prosumer"—a professional software package with only the entry-level or minimum features expected by a professional software editor. The leading prosumer video editing software package is Adobe Premiere.

These software packages pack all the power of the professional systems used to edit broadcast TV and movies. They include basic functions of video editing and even the neophyte will find that using these packages is no more difficult than the intermediate packages, but they contain much more powerful features and many more transitions and effects. They cost from $500 to $850 if purchased individually. Adobe Premiere comes bundled with many of the $500+ FireWire cards. Imagine, a FireWire card *and* a professional editing package for less than the software alone. This is a bargain!

Notice that the Premiere user interface shown in Figure 7-6 resembles that of the MainActor screens. What you learned using MainActor easily transfers to Premiere. You will also notice that there are many more choices on the screen.

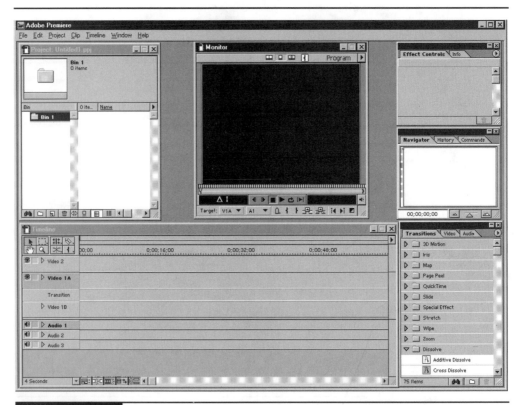

FIGURE 7-6   Premiere main edit screen

Premiere offers features beyond the basics of intermediate packages. These features include:

- Batch capturing of clips. Batch capturing allows the editor to create a list of edits with their in and out points and timing. The computer automatically does the edits while the editor sips coffee.

- A/B and single timeline professional interface.

- Two-screen monitoring, extensive title functions.

- Countless effects as well as video and audio filters for the widest possible range of design options. Transparency effects for compositing.

- Fully integrated capture, editing, titling, export, and so on.

- Professional audio editing.

If you are planning to do anything but the most basic and occasional editing and can afford to get Adobe Premiere or a similar package such as Razor Cut Pro or LT, you will not be sorry. Their ease of use and ultimate power will serve and inspire you to undertake much more sophisticated video projects. If you can't afford a package like these, you can still make very professional, but not so complex, finished videos. The technical quality of the finished video is exactly the same with all three packages. The difference is the number of special effects and features of the packages, and their ease of use.

Summary

In this chapter we have gone through the process of selecting a FireWire connection and an appropriate video editing software package to use with it. The choices ranged from simple editing utilities that can be purchased for less than $100 to prosumer editing software that will allow you to create your own *Blair Witch Project*. The joy of shooting video is only the first step. The editing and processing of the video into real productions, even if simple, is truly a satisfying task and one that will be appreciated by your friends and family. In the balance of the book, you'll learn how to capture and edit video and to create finished video for TV and computer use.

Chapter 8

Transfer Your Video Clips and Edit Them

How to...

- Organize your video scenes and shots
- Understand the capture process
- Connect your camcorder to your computer for capturing video
- Capture your clips and bring them into the editing software environment

Once you have installed your FireWire connector in your computer, you will want to connect your camcorder to the computer and transfer video to the hard drive so you can manipulate and edit it. This process, a holdover from transferring analog video into computers, is commonly called video capture. In order to begin capturing, you will need, in addition to the FireWire hardware and cables, a piece of software that can control the camcorder and enable the transfer process. In the case of digital video, however, the video is already in a computer digital format, while still stored on the camcorder tape. Unlike analog video, it doesn't need to be converted, but only transferred from the digital tape to the digital hard drive.

The DV-Capture software utility will enable this process. Once the video clips are transferred from the camcorder to the computer using the DV-Capture utility, they can be added to the video editing software's trim and timeline programs for assembly. Here you will be adding titles, transitions, and other effects, and ultimately creating a finished video program containing all these elements. Once the final program is created, it will be transferred back (if you want it to be used on tape) to the DV camcorder using a DV-Out utility program. But let's not get ahead of ourselves; I will cover this in Chapter 14. Here I will focus on the creation of the final program.

Making a Scene About Your Video Project

Every video is made up of scenes. A scene is a section of your content. We're all quite familiar with this filmmaking term where a small unit or portion of the overall story is captured on film. It is the same with video. If you are shooting a series of pieces of animal footage from your vacation in Yellowstone, for example, each separate bit of taping with a beginning and an end would be considered a scene. When you move to another type of action, different set of characters, different location, different time, or different emphasis, you have a different scene.

When you play it back, the tape moves from one subject or scene to the next. The problem with using this tape as your finished video is that you rarely can start and stop the camcorder at exactly the right time, recording only the material you want in your final video. In addition, you often will want to eliminate or rearrange the scenes that you have recorded or add scenes that were recorded on different tapes. Finally, you will probably want to add titles and transition effects to make your video look professional. The first step in this process is to transfer (capture) the individual scenes from the tape or tapes you have recorded onto your computer so that you can edit, rearrange, or add titles and effects to create a final project. The scenes that you transfer are called *clips* and are transferred using a utility software program provided by your editing software publisher. This software can control your camcorder and computer and enable the connection's selections and transfers. This utility can be a separate program that is used before you open your main video editing program or it can be integrated into the program itself. The two editing packages that we will consider in this book are MainActor by Dazzle and Premiere by Adobe. MainActor provides a separate utility called DV-Capture and Adobe Premiere has the capture utility built into the main editing program's menu. Both programs, in basic mode, accomplish the same task in similar ways, though the Premiere program has a few extra features such as batch capture, which we will discuss later.

Organizing Your Project

Navigating through a tape and trying to decide what scenes to capture is difficult under the best conditions, but without a systematic procedure it can become a frustrating nightmare. Most videographers and video editors create shooting schedules that consist of lists of the proposed scenes and the shots that make up those scenes. The shooting schedule is just that— a list of the scenes and shots you create during the shooting of your project. The shooting schedule is usually based on the script of planned shots initiated at the time you shoot and can even include begin and end times of the scenes on the list. (This process in described in more detail in Chapter 6.) One of the advantages of creating a shooting schedule with scene and shot lists is that it allows you to think ahead to your finished video. When you are creating your scene list during the shooting process, you can stop and preview the tape and determine which scenes you plan to capture for use in your final video. At this point you can make a note of what you wish to name each scene file. This can be noted on the scene/shot lists or you can create a separate edit list for each tape you are capturing from and reference it back to your shooting schedule or script. If you use the shooting schedule without an additional edit list, note any changes made while you are capturing. You will find it much easier to recall what each clip is to be used for when you are doing your timeline editing.

   The way I do it is to create shot lists based on scenes. A scene is generally made up of multiple shots. For instance, in a birthday video one of the scenes might be the cutting of the cake. The shots that make up that scene might be the birthday boy's face, the lighting of the candles, the guests waiting for cake, and the cake being cut.

The DV-Capture Utility Program

This section will guide you through the process of moving your video clips from your camcorder or other DV playback device to your computer for editing. I used the Dazzle MainActor DV-Capture utility program to demonstrate the process, but the process is similar in all the other edit and capture programs I have used. Refer to the manual for your particular software if it is not the Dazzle program.

Step One: Understanding Your DV-Capture Software

Using MainActor DV-Capture (see Figure 8-1), the digital video capture utility program provided with MainActor, you will control the digital camcorder and capture or copy video scenes from the camcorder to the computer's hard drive. When you are using MainActor, this utility will be installed along with the other elements of the MainActor program.

FIGURE 8-1   MainActor DV-Capture screen

The DV-Capture Functions

The menu options for the DV-Capture utility are as follows:

File   This menu lets you save your captured clip and assign it a file name.

Capture Device   This menu lets you select the DV device from which you want to capture. It will automatically list all of the DV devices, including camcorders and playback decks, that are attached to the FireWire connections of your computer and are turned on. A second menu option will allow you to reset the options if you have added or subtracted devices.

Capture   This menu offers two choices: Record and Stop. Record begins the capture process from the camcorder, beginning capture where the tape is set. Stop stops the capture process.

Settings   This menu lets you select the source and display features of the utility.

Help   This menu provides online help.

The Remote Control Panel

This panel (see Figure 8-2) functions if your device supports remote control over the FireWire connection.

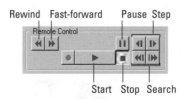

FIGURE 8-2   The Remote Control panel

 *Some digital camcorders or DV players may not support remote control operation when connected to the computer with the FireWire cable.*

The buttons are used as follows:

Rewind and Fast-forward   These buttons rewind and fast-forward the video.

Pause   This button is used to pause the video.

Step   These buttons step the video forward or backward by a single frame.

Search   Just like on your VCR, these buttons search the video forward or backward at faster than normal speed so you can look for specific points in the video.

Start and Stop   These buttons start and stop playing the video.

The Capture Control Panel

These buttons, shown in the illustration that follows, control the capture process itself. The left button begins the capture, and the right button stops it. You can also begin capture by pressing CTRL-R and stop capture by pressing CTRL-S.

The DV video is copied directly from the camcorder tape to your hard drive in real time. In other words, it will take about a minute to copy a minute's worth of video to the hard drive.

The Capture File Name Display

The capture file name display, shown below, shows the path and file name the captured clip will use. If you click the disk icon on the left of the display, you may browse and change the path and save the file name.

Time Code Display

The time code display, shown in the illustration below, shows the DV tape's time code so that you can find the place on the tape for the "begin" and "end" capture commands.

Timecode 00:00:00.00

Step Two: Capturing a Scene with DV-Capture

Now that you have been introduced to the control panel, the next step is to actually capture video and transfer it from your camcorder into your computer. The following steps make this easy:

1. Attach the FireWire DV cable to your camcorder or other device and to the FireWire connector on your computer.

2. Turn on the camcorder and wait a few seconds for the computer to recognize the DV source.

3. Launch the DV-Capture utility program.

4. Select the DV-Source you want to capture from (see Figure 8-3). If you have only one device plugged into the computer it will be the default.

5. Select the file name you want to give the scene you are capturing (see Figure 8-4).

NOTE *One second of captured video takes about 4 megabytes of hard disk space, so make sure that your hard drive has enough space for the size of the clips you plan to capture. Windows 98 will only allow the capture of a file of 4 gigabytes or about 19 minutes of video. If you want to save more total footage than that, it will need to be broken into smaller scenes. Windows NT 4.0 and above has no file size limit if you have installed it with the NTFS or FAT 32 file system instead of the 16-bit FAT file system. See your Windows manual for information about these systems and their installation.*

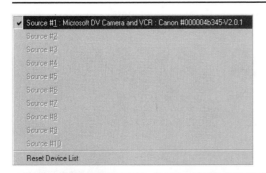

FIGURE 8-3   DV-Source device list

FIGURE 8-4   Save As dialog box

8

6. If your device supports remote control, click Play on the Remote Control panel to play the DV tape and to select the start point for your scene. If your device doesn't support remote control, you will have to control it from the buttons on the device. When the scene begins, it will play on your computer screen (see Figure 8-5).

7. When you have reached the start point on the tape for your scene, click Record in the Capture control panel to begin capture. Begin the capture process a second or two before the beginning of the scene and let it play a second or two beyond the end. You can trim the beginning and end later during the editing process to eliminate the excess. There will be no preview on the computer screen when capture is taking place.

TIP   *You can watch the playback on your camcorder viewfinder. I have attached a small 9-inch video monitor with AV connectors to the analog video output of my camcorder so I can view it next to my computer monitor during capture.*

FIGURE 8-5   DV playback screen

8. When the end of the scene arrives, click Stop to end capture.

9. You may continue to capture as many clips as you need for your project.

When you have finished the capture session, you can open the MainActor Sequencer to edit and assemble your video project from the captured scenes. This process will be covered in the next few chapters.

Batch Capturing Clips

Batch capture is a great feature if you are capturing a large number of clips or need to recapture clips for rebuilding a project later. Capture software that has this feature allows you to create a list of start and end times for your DV tape, followed by a simple click to begin capture. The computer will then fast-forward to the next clip, capture and save the clip to the prearranged file name, and move on to the next clip, while you go off and have coffee. MainActor's DV-Capture utility doesn't support batch capturing, but other editing programs have capturing utilities that do. They may also be purchased separately. Check with your software store or video websites for availability.

Summary

In this chapter, I have covered the basics of connecting your DV camcorder or device to your FireWire-equipped computer so that the video you have shot can be edited and processed. I have also covered the basics of capturing or transferring your digital video to your computer and making it ready for the editing process. You are now ready to begin learning about trimming, editing, and other processes as we move ahead.

8

Chapter 9

Create an Editing Timeline

How to...

■ Prepare your clips for insertion into the editing timeline

■ Insert a captured video clip into the MainActor Sequencer timeline

■ Drop a video transition onto the timeline

■ Drop a video effect onto the timeline

■ Add a second clip to the timeline

■ Export the video as an AVI or MPEG file

In this chapter you will take all the little pieces of video you have shot and, using your editing software, create a seamless section of video. In some ways this is like a quilter creating a coherent pattern from dozens of scraps of fabric. Here you will learn to trim away what is not needed and connect the video clips together, bringing your story to life.

Using the MainActor Video Editor

Once you have selected and moved your individual clips from your digital videotape to your computer hard drive, you will want to preprocess the clips to assure that they begin and end on the exact frame you want. This will eliminate bits of video that you don't want in your final video. You also might want to break a clip into more than one part or remove pieces from within the clip itself. With some programs, such as Adobe Premiere, these operations can be done with the editing software itself, as well as more technical and advanced processes such as changing the clip palette, the set of colors that are used within it, or the frame size or file type. With Dazzle MainActor, these processes need to be done in a separate program called the Video Editor (VE). MainActor VE is a video/animation composing and processing utility. It allows you to load, edit, play, and save or convert video, animations, still pictures, and sounds of any size and of various formats. I will open and trim the beginning and end of a sample clip to give you an idea of what you can do.

To begin, click the MainActor VE editor icon from the Windows Program Manager or the directory it is stored in, to open the editor utility, as shown in Figure 9-1.

Next, from the main menu, select File | Open and select the clip you want to edit. The editor creates a project for you to work on, as shown in Figure 9-2. A project is either an animation/ picture or one or more pictures of the same format. All pictures of the same format will always be grouped together in one project, and this project is named after the picture format. You should have imported one or more clips from your digital videotape in Chapter 7 (if not, go back and read that chapter now). The file you select will show up in the project area on the left side of the editor window. The frames of the currently selected project are displayed in the Frame list. Each project has its own pop-up menu, which when double-clicked will play back the project. When you select the file or range of still images you want to load, click OK and the file will be loaded into the editor as a project.

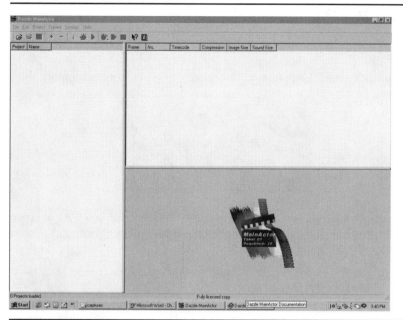

FIGURE 9-1   MainActor editor screen

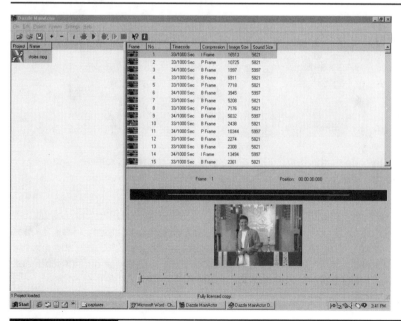

FIGURE 9-2   A selected project

Trimming a Video Clip

Now that you have selected a clip to process, let's trim 100 frames from the beginning just to understand the process.

1. From the Edit menu, select Range. This opens the Select Range dialog box shown here:

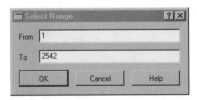

    This dialog box allows you to select the range of frames that you want to process. At this point you have not determined what the process will be. Notice that your sample has a total of 2542 frames.

2. Select the range 100 to 2542. This determines that whatever process you select next will be applied to this range of frames.

3. Save the range as another file by selecting File | Save. This will open the Save Window dialog box, as shown here:

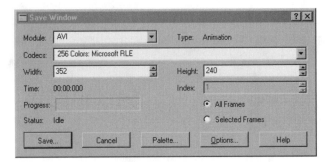

    To trim the clip of the first 100 frames, you will save these selected frames as a new file and that will become the edited clip less the first 100 frames.

4. To select the kind of file format you want to save your clip in, click Options. The Options dialog box opens from which you can choose a file format. The choices include AVI (Microsoft's Audio Visual format), MPEG, and a number of other choices. In most cases, when you are trimming clips for use in the Sequencer timeline you should choose the same type and size of file as your original clip. This will be the default setting of the clip when you select Save.

5. Click OK. You will be returned to the Save Window dialog box.

6. Now that you've made all of your selections, click Save. The Select File dialog box appears, as shown in Figure 9-3. Choose a directory and a file name and click Save.

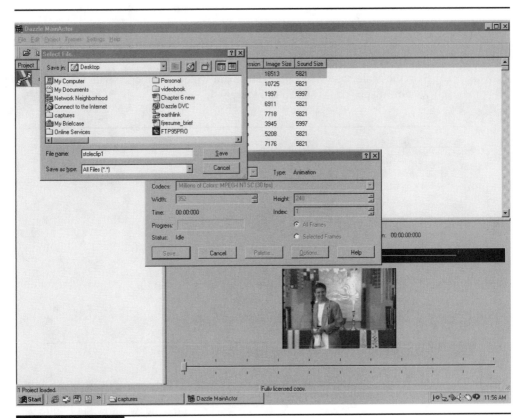

FIGURE 9-3   Select File dialog box

The clip is trimmed and ready to be inserted in the MainActor Sequencer timeline.

Editing a Simple Video with MainActor Sequencer

Once you have gathered together all of your clips and have done your preliminary trimming and processing on a clip-by-clip basis, you are ready to assemble the clips into a full video program and add titles, effects, and transitions. In this section, I will cover the basics of the video editing software package MainActor Sequencer and show you how to import clips and add a simple transition. The process in MainActor is similar to Adobe Premiere and other editing packages. I will cover these topics in greater depth later in the book.

Inserting a Video Clip

The first step in creating your video is to insert a video clip into the timeline. The video clip will become the first element of your video. You will also add a second clip, a transition, and a video effect.

First, you need to start the Main Actor Sequencer and select the appropriate profile for your project.

1. Start the MainActor Sequencer. The main window will open with the Video Profile window displayed, as shown in Figure 9-4.

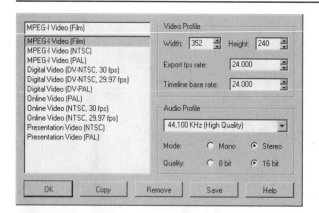

FIGURE 9-4   MainActor Video Profile window

2. Highlight Digital Video (DV-NTSC, 30 fps) from the Video Profile menu and click OK. You will then see the MainActor Sequencer main window, shown in Figure 9-5. You can stretch and drag the various windows of the Sequencer into any size or order you want.

The features of the MainActor Sequencer window include the menu bar and other windows used to process and edit the video projects. The menu bar includes File, Edit, Object, View, and Help. The first toolbar has a series of icons that access the primary functions of the Sequencer. The editing functions include the following:

New project   This button creates a new project.

Open project   This button opens an existing project.

Save project   This button saves the active project.

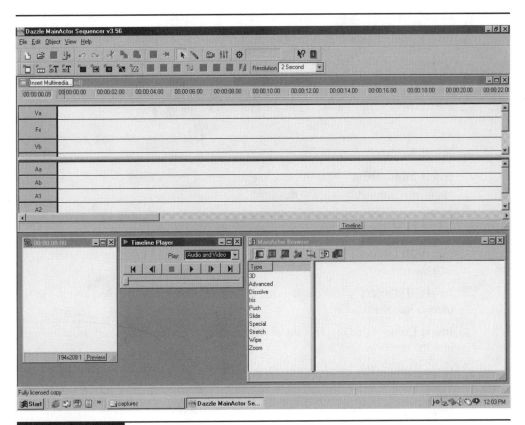

FIGURE 9-5   MainActor Sequencer main window

Export   This button exports objects and projects.

Undo the last command   This button will undo the last command used.

Redo the last command   This button will redo the last command used.

Cut to clipboard   This button cuts the selected object from the timeline and stores it on the Windows clipboard.

Copy to clipboard   This button copies the selected object to the Windows clipboard.

Paste from clipboard   This button pastes the current object in the Windows clipboard to the timeline.

Reverse object   This button indicates to the program to play the selected object (audio, video, animation) backwards.

Pin object   This button places an object in the timeline in a fixed position.

Edit mode   This activates the Edit mode.

Cut mode   This allows the trimming or cutting of an object in the timeline.

Capture video   This button opens the video capture function.

Preferences   This button opens the Preferences dialog box.

The following are Sequencer functions that are accessed by selecting a menu item, which opens a dialog box:

Insert Multimedia   This displays the dialog box that is used to select and insert video, audio, and animation into the timeline.

Insert Pictures   This opens the dialog box that is used to select and insert still images into the timeline.

Insert 2D Object   This opens the dialog box that is used to select and insert 2D objects into the timeline.

Insert 3D Object   This displays the dialog box that is used to select and insert 3D objects into the timeline.

Insert Color   This displays the dialog box that is used to select and insert colored frames into the timeline.

Object Info   This opens a dialog box that displays the properties of the selected object in the timeline.

Object Video Effects   This opens the Video Effects dialog box.

Object Video Paths   This opens the dialog box to choose the video paths to the objects stored on your computer.

Object Audio Effects   This displays the Audio Configuration dialog box.

Object Settings   This opens the Settings dialog box for the type of object you have selected.

Object Overlay Settings   This displays the dialog box that controls the transparency and other properties of an object being overlaid on another object.

Object Video Settings   This opens the dialog box that allows you to change video timeline settings for an object.

Split Object   This command splits the audio and video tracks into two separate objects for individual processing and editing.

Resolution   This command changes the divisions of time used to display the objects in the timeline.

Below the toolbars is the timeline itself. This series of horizontal tracks is where you insert and edit video, audio, animation, and other objects and create a program from the various objects and clips. The window is made up of a series of tracks. Va and Vb are the two tracks for inserting video, animation, and still images. Fx is the track used for effects and transitions. Aa and Ab are used for audio tracks. A1, A2, A3, and so on to A96 are used for overlaying effects on the primary Va/Fx/Vb tracks. You might insert scrolling titles over the opening sequence using this function, for example. You could also set transparency to govern the amount of image that shows through the title or other overlay.

Below the timeline are the video preview window, which displays your video clips and project, and the Timeline Player that controls playback. The last window is the menu of objects, transitions, and effects.

Now let's insert your video clips and other files.

1. In the toolbar, click Insert Multimedia, located on the second row on the far left (see Figure 9-5). Import Dialog will appear, as shown here:

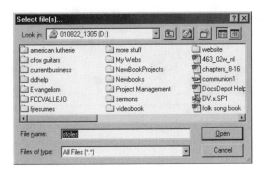

2. Click OK. The Select Files dialog box will appear, as shown below, in which you can select a video clip that you have captured from your camcorder. You can refer to Chapter 7 where this was discussed.

3. Select the file you want to import and click Open.

4. Move the cursor to the Va track. The cursor will change to a small film icon attached in the lower right indicating that you are ready to insert your video clip. There will also be a gray rectangle in the track that indicates the length of the video clip, as shown in Figure 9-6.

5. Drag the clip with the cursor along the Va track until the timer reading is 00:00:00:01. This indicates that you are at the first frame of the video clip that you are creating.

6. Click to place the video clip on the timeline.

7. Unless it is a very short clip, you will want to change the resolution of the timeline from the default two seconds to a larger number. On the toolbar you will find a pulldown menu labeled Resolution that will accomplish this. Let's leave it at two seconds for the time being.

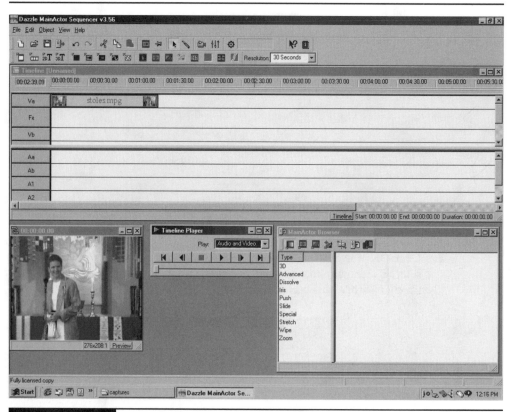

FIGURE 9-6   Inserting video clip into the Va track

Inserting a Second Video Clip

This sequence shows the process for inserting a second video clip in your timeline.

1. Change the resolution, see step 7 in the previous section, to 30 seconds so that you can see a longer duration of the timeline.

2. Now select a second clip from your directory as you did earlier and insert it in the Vb track following the first clip, as shown in Figure 9-7.

3. You can continue to add clips and transitions to complete your project. You may insert clips in either track Va or Vb. If you use Va, the video will play from the first to the second clip without transition.

4. If you want to insert a transition, alternate the use of Va and Vb with a transition between the two clips allowing the transition to play from the clip in Va to the clip in Vb and so forth. I will cover titles and more advanced transitions in Chapter 10.

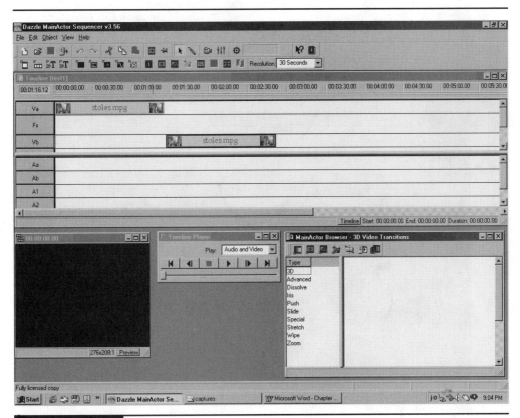

FIGURE 9-7   Inserting a second video clip into your timeline

Inserting a Transition into Your Project

Next you will add a transition to the beginning of the video clip.

1. In the lower-right corner of your main window, you will see the Browser window. From the selection of Transition icons, find the one that looks like a swinging door opening to reveal the video clip you just inserted.

2. In the Browser window, make sure that the Video Transitions button on the far left of the toolbar is selected and then select 3D in the Type list, as shown in Figure 9-8.

3. The transition you will use is the one in the upper-left corner that looks like vertical doors swinging back and forth. Click and drag it to the Fx track in the timeline. Release the mouse button and the cursor will change to indicate that you are positioning this transition.

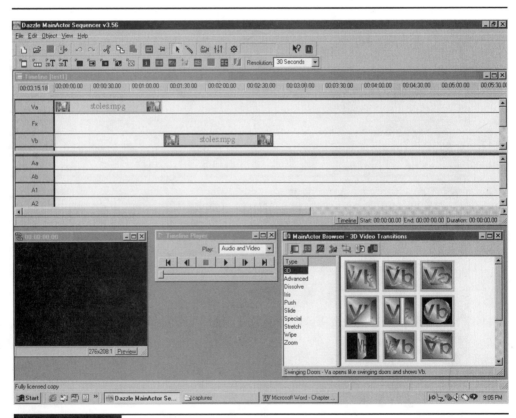

FIGURE 9-8   Selecting the 3D video transition from the Browser window

4. Move the cursor so that the transition is at the end of the first clip of the timeline in the Fx track and click to place the transition.

5. Move the cursor over the transition so that the cursor changes to horizontal arrows indicating that you can change the length of the transition. The duration of the transition will be displayed in the timeline window. It will read, "Start: 00:xx.xx.xx End: 00.xx.xx.xx Duration: 00.00.03.00," indicating 3 seconds. The xx will vary depending on the length of the clips you inserted.

6. Drag the clip so that about half overlaps the end of clip 1 in the Va track and half overlaps the beginning of clip 2 in the Vb track.

7. Click and drag the edge of the transition until it extends to three seconds on the timeline. The timeline should now look like the one shown in Figure 9-9.

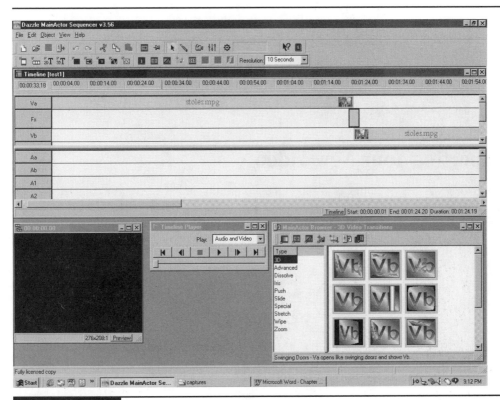

FIGURE 9-9   Transition inserted in the timeline

8. If you right-click on the transition in Fx you will open a menu that includes information and settings. Open the settings dialog box for the effect shown here:

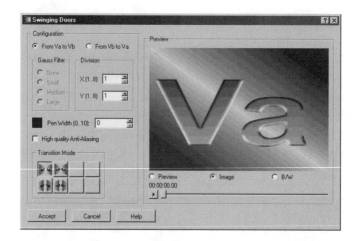

This dialog box lets you determine the effects of the transition. In this case you will want to choose whether the transition moves from the Va to the Vb or vice versa. Select the Va to Vb choice.

Exporting Your New Video

The final step of editing is to export the project to your hard drive as a completed video. The default choice for DV would be in DV format so that you can print it back to DV videotape. Other choices might be MPEG for distribution on CD-ROM or RealVideo for Internet distribution.

1. Click the Export button on the timeline menu bar to open the Export dialog box shown here:

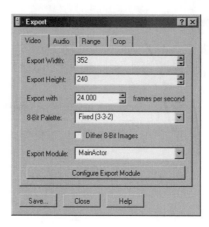

2.   Click Configure Export Module and select the format for AVI or MPEG if you want to experiment. We will select AVI instead of MPEG. Click Accept and you will return to the Export dialog box.

3.   Click Save and the Select File dialog box will open, as shown here:

4.   Save the file as you would in any Windows application.

You have created, edited, and exported your first video program.

Summary

This chapter gave you a good overview of the main timeline editor of MainActor, which is typical of video editing software packages. You have learned to insert video clips, transitions, and effects and to save the resulting edits into a finished video project file. These techniques are the basics of editing and will constitute the majority of the editing functions you will be required to do. I will cover the details of effects, transitions, titles, and more in subsequent chapters. I will also walk through entire sample projects in the final section of the book.

Chapter 10

Add Transitions and Effects to Your Timeline

How to...

- Insert transitions into the timeline
- Review types and applications of transitions
- Insert effects into the timeline
- Review types and applications of effects

Video transitions are computer functions that are used to smoothly change from one video clip or still graphic to another. Most video editing is accomplished by cuts, where one clip is butted against the next without any special transition effect. In some cases, such as in titles and between major scenes or if a special emphasis is desired, a video editor may choose to add a more dynamic transition. Common transitions include dissolves, where one scene dissolves and the next one appears through it, and swinging doors, where a door appears to open in the middle of one scene and the next scene appears thorough it.

Video effects are special processes that use color, light, and so on to add changes to individual video clips, still images, or animation. They are not used to change from one video object to another. One familiar video effect is the *morph*, where one face or object seems to transform into another. Transitions can also contain or be based on special effects, but they are most commonly used to move from one clip to another.

All editing software comes with at least a minimal set of built-in transitions and effects that can be used in your video. The more advanced programs like MainActor have a diverse palette of transitions and effects. Top-level packages such as Premiere have transitions and effects built in as well as the ability to work with additional transitions and effects purchased from third-party software makers. These plug-ins will work as if they were originally part of the program, giving you a much wider choice of transitions and effects to choose from in your video editing.

Creating Effective Transitions

Just because you can do something doesn't mean you should. Cutesy transitions are sometimes appropriate, but not often. Think about the *Batman* TV show—scenes frequently folded up or exploded outward, usually accompanied by a cartoon word in a bubble—Pow! Bang! It's hard to imagine these transitions being useful or effective in a video production with any serious content. However, a video of a child's birthday party could be really fun with some of these transition effects.

In video, as in film, the cut and the dissolve are the two transitions most frequently used. These can be used in some very creative ways. One of the best transitions I have seen recently occurred in an episode of the TV show *The West Wing*. The main character, now the President of the United States, was remembering scenes from his youth involving a woman who would become his secretary in the White House. The secretary had been recently killed in an automobile accident. The scenes went back and forth between the past and the present as he recalled her influence on his life. The most effective scene occurred when the young man finished a conversation with her and walked away slowly, disappearing behind a pillar at the edge of the scene. The camera slowly panned past the pillar, and on the other side, the adult President is in the National Cathedral at the funeral. The actual edit is a simple cut, but the two pieces of film are so carefully placed together that the scenes merge seamlessly, bringing the past into the present. Such transitions do not have to be fussy or fancy to be effective—just done with care and attention to detail.

Applying Video Transitions Using Sequencer

I will use the Peel transition to illustrate the process, but all transitions are created in a similar way. To add selected transitions to your video, use the following steps:

1. Open the MainActor Sequencer. The Sequencer main screen is shown in Figure 10-1.

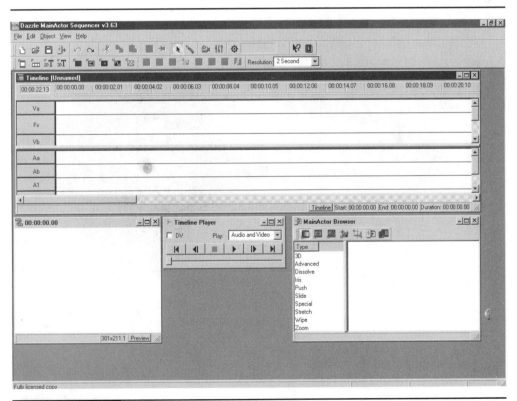

FIGURE 10-1   Sequencer main screen

2. Right-click the timeline and select Insert Multimedia. Select a bitmap file from the File dialog box.

3. When you select the file, the cursor will change into a bitmap. If you drag the bitmap onto the timeline, the bitmap will change to indicate where you should place it. To place the bitmap on the timeline, place the cursor at the beginning of the Va track and click (see Figure 10-2).

FIGURE 10-2   Inserting a bitmap in the timeline

4. Move the cursor over the right side of the bitmap until the cursor becomes a horizontal arrow. Click and hold to drag the frame of the bitmap until it extends to the 00:00:05 mark on the timeline, and then release the mouse button.

5. Add a second bitmap in the Vb track using this same procedure.

6. Go to the Browser window and select the Transitions button. Click 3D, select the Peel transition from the top middle line of icons, and drag it to the Fx track in the timeline. When the cursor changes to indicate that you may place the transition

on the timeline, move it to the beginning of the timeline and click to leave it there, as shown in Figure 10-3.

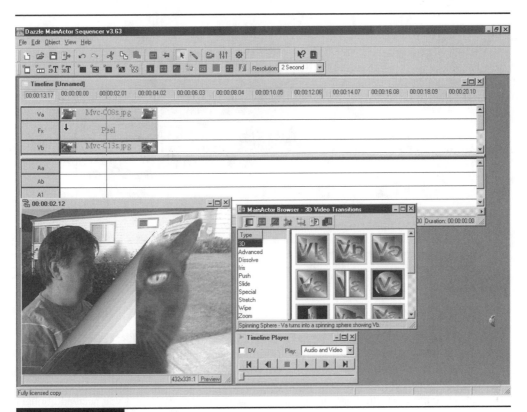

FIGURE 10-3   Inserting a transition from the Browser window

7. Move the cursor to the left edge of the transition object until the horizontal arrow appears, then click and drag the edge of the transition object until it matches the edges of the bitmap objects at about the 00:00:05 mark on the timeline.

8. You can change the settings of your transition. To do so, double-click the transition object. This will open the settings window for the transition. (Figure 10-4 shows the settings window for the Peel transition.) This window controls how a particular video transition is applied. Depending on the type of transition, various parts of this window may be disabled if they do not apply to the current transition. Make sure that the configuration is checked for From Va To Vb. This indicates that the transition moves from the object in track Va to the object in track Vb.

9. You can then add more objects, transitions, effects, and titles to your project. Once you have finished, save the project or export it to video.

FIGURE 10-4   Transition settings window

Transitions Available in MainActor Sequencer

Most video editing packages above the entry level will contain stock video transitions. As described earlier, both MainActor and Adobe Premiere include a fairly extensive set. You can also purchase transition plug-ins from third-party software vendors if you need additional effects for Premiere and other more advanced editing software packages. MainActor doesn't support third-party plug-ins. You'll see how built-in transitions are used with MainActor and learn about the transitions that are typical of MainActor, Adobe Premiere, and other advanced software packages in the sections that follow.

3D Transitions

3D transitions are created in three dimensions—that is, they have depth in addition to horizontal and vertical characteristics. These include:

Swinging Door   Opens the Va track like a swinging door to reveal the Vb track (see Figure 10-5). This is a good standard transition for scene changes.

Peel   Peels back the Va track to reveal the Vb track (see Figure 10-6). This is a good standard transition for scene changes.

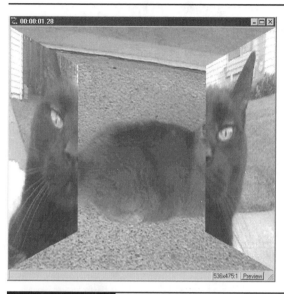

FIGURE 10-5   Swinging Door transition

FIGURE 10-6   Peel transition

10

Roll   Rolls back the Va track to reveal the Vb track (see Figure 10-7).

Spinning Sphere   Turns the Va track into a spinning sphere that reveals the Vb track (see Figure 10-8). This works best for action endings.

Spin Away   Rotates the Va track at about the center and reveals the Vb track (see Figure 10-9). This is a good ending transition.

Swing   Presents a single door opening the Va track to the Vb track (see Figure 10-10).

Transparent Roll   A roll that has a transparent background (see Figure 10-11). This is a good standard transition for inter-scene clips.

Venetian Blind   Alternates slices of the Va track with the Vb track, giving the effect of Venetian blinds opening or closing (see Figure 10-12). This works well on a sequence of title slides.

FIGURE 10-7   Roll transition

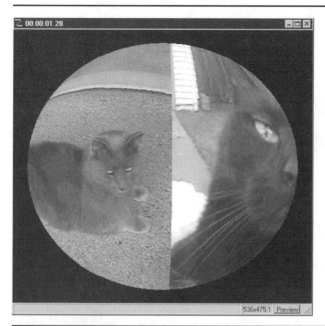

FIGURE 10-8   Spinning Sphere transition

FIGURE 10-9   Spin Away transition

10

FIGURE 10-10   Swing transition

FIGURE 10-11   Transparent Roll transition

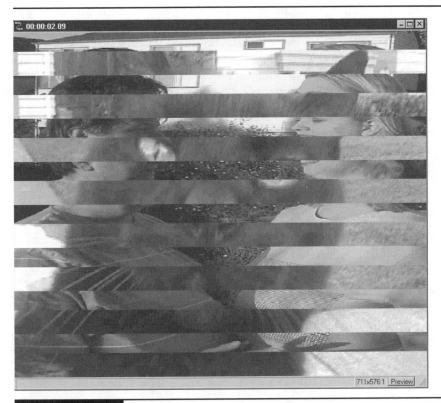

FIGURE 10-12   Venetian Blind transition

Advanced Transitions

These advanced transitions are more sophisticated and more difficult to set up and utilize. They are meant for special emphasis or for adding action and excitement to an otherwise ordinary video sequence.

Morph

The Morph transition lets you move and blend the features of one image into specified features of another image. It can be used to create high-quality morphs from either Va to Vb or Vb to Va. Morphs are defined by drawing control lines on the objects chosen to morph in the Va and Vb tracks. The morph control lines are visible in both Va and Vb windows.

TIP

Morphing is a spectacular transition that requires practice and restraint in its application in finished video projects. It has been overdone in movies and television.

To create morphs, click the beginning, or inpoint, of a control line in the control window, as shown in Figure 10-13, and drag the currently selected control line, or end points, in both windows separately. This defines how a line corresponds to the equivalent line in Va/Vb. A line may join a mouth, eyebrow, or nose. Increasing the line length controls the amount of morphing effect; the longer the line, the greater effect of the morphing for this line. You can place and control up to 100 control lines by using the Line numeric entry field or by selecting the current line by clicking with the cursor.

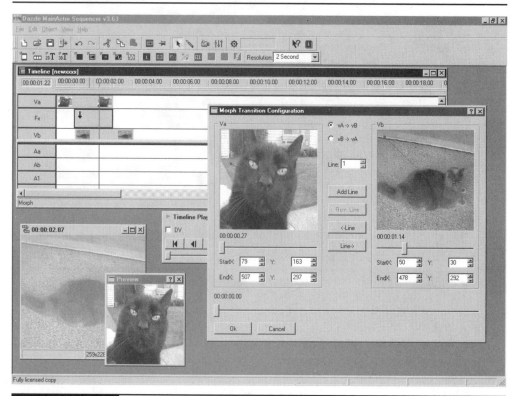

FIGURE 10-13   Morph window

Additional lines are added by clicking the Add Line button. Remove lines by clicking the Rem. Line button. The Line-> and <-Line buttons copy the line settings from the left window (Va) to the right window (Vb) or vice versa. The line and settings for a mouth or nose in the Va window can be copied into the Vb window and modified for effect.

Dissolve

There are four Dissolve transitions:

Dither   Dissolves the Va track into the Vb track with a dithering (or smoothing) effect.

Additive   Smoothly dissolves the Va track into the Vb track (see Figure 10-14).

Mosaic   Dissolves the Va track into the Vb track with random large blocks.

Random   Dissolves the Va track into the Vb track by randomly changing the pixels.

FIGURE 10-14   Additive dissolve transition

 The dissolve is one of the most commonly used and most effective transitions. Try a variety of settings and durations on a sample video clip until you understand its potential fully.

10

Iris

There are several transitions that utilize openings to move from one clip to another. This works rather like the iris of your eye that opens and shrinks depending on the light. In fact, these transitions are called Iris transitions. There are five Iris transitions: Cross, Diamond, Round, Square, and Star. As you might suspect, each uses an opening shaped like a cross, diamond, and so on to move from the Va track to the Vb track. Figure 10-15 shows the Cross Iris transition.

FIGURE 10-15   Cross Iris transition

Push

The Push transition includes Band and Push variations. The Band transition uses horizontal bands pushed to the side to transition from the Va track to the Vb track. The basic Push transition pushes the Va track horizontally to reveal the Vb track, as shown in Figure 10-16.

Slide

There are ten variations of the Slide transition:

Band   Uses horizontal bands to slide the Va track to the Vb track.

Cross Split Splits the Va track into four rectangles and slides them out through the corners to reveal the Vb track.

Flying Carpet Uses a magic carpet–like wave effect to transition from the Va track to the Vb track.

Inset Slides from one corner to the opposite corner to move from the Va track to the Vb track.

Rotate Rotates around one corner to move from the Va track to the Vb track (see Figure 10-17).

Sliding Doors Opens like sliding doors to move from the Va track to the Vb track.

Sliding Boxes Uses sliding vertical strips to move from the Va track to the Vb track.

Basic Slide Slides the Vb track from one side onto the Va track.

Sliding Quarters The four corners of the Va track slide away one by one to reveal Vb.

Swap Slides the Va and Vb tracks horizontally and reverses at the halfway point so that the Vb track appears to jump in front of the Va track.

10

FIGURE 10-16   Basic Push transition

| FIGURE 10-17 | Rotate Slide transition |

Special

There are six Special transitions:

Build   Resembles a rising wall covering the Va track with the Vb track.

Clock   Sweeps from the Va track to the Vb track, resembling the movement of the hands of a clock.

Shredder   Shreds the Va track horizontally, revealing the Vb track (see Figure 10-18).

Spiral   Uses a spiraling block to overlay the Vb track on the Va track.

Pinwheel   Reveals the Vb track with the Va track disappearing behind rotating vanes of a pinwheel effect.

Windscreen Wiper   Wipes the Va track using a rotation centered at the top middle, revealing the Vb track.

TIP   *Use these kinds of transitions sparingly. They're fun, but they can overwhelm a project when used too often.*

00:00:02.12

711x576:1 Preview

FIGURE 10-18 Special Shredder transition

10

Stretch

There are two Stretch transitions: Cross and Cube. The Cross transition is a 2D transition that squeezes the Va track horizontally to reveal the Vb track (see Figure 10-19). The Cube transition is a 3D transition that squeezes the Va track horizontally, revealing the Vb track.

Wipe

There are three Wipe transitions: Band, Checker, and the basic Wipe. The Band transition uses two horizontal bands to cover the Va track with the Vb track. Checker transitions, as shown in Figure 10-20, from the Va track to the Vb track using a checkerboard pattern. The basic Wipe transition horizontally covers the Va track with the Vb track.

TIP

Checker Wipe is a good transition to use for titles and title slides.

FIGURE 10-19   Cross Stretch transition

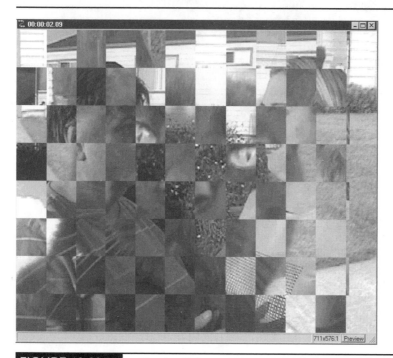

FIGURE 10-20   Checker Wipe transition

Zoom

There are two Zoom transitions: the Corner and the Center. The Corner Zoom transition begins in the corner and expands the Vb track to cover the Va track. The Center Zoom transition starts in the center and expands to cover the Va track with the Vb track (see Figure 10-21).

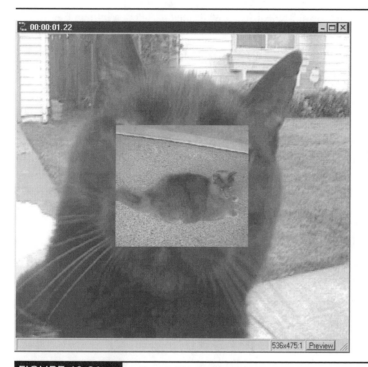

00:00:01.22

536x475:1 Preview

FIGURE 10-21   Center Zoom transition

10

Using Video Effects in Your Video Production

Effects are used to change the appearance of a graphic, bitmap, or video clip. They can be used within the editing package to change the appearance of a clip before it is added to the program or to add a dynamic effect to the video clip itself. Color or lighting balancing are examples of a static effect where the changes to the video are made during the editing process, and the viewer only sees the video after the changes are made. Morphing is a good example of a dynamic effect where the viewer actually sees the change taking place.

How to Apply Special Effects to Your Video Clips

I will use the Swirl effect to demonstrate the addition of an effect to your project, but all effects are created in a similar way. To add effects to your video:

1. Open the MainActor Sequencer.

2. Right-click the timeline and select Insert Multimedia. Select a bitmap file from the File dialog box.

3. When you open the file, the cursor will change. If you drag the cursor onto the timeline, it will indicate where you should place the bitmap, as shown in Figure 10-22. To place the bitmap on the timeline, place the cursor at the beginning of the Va track and click.

4. Move the cursor over the right side of the bitmap until the cursor becomes a horizontal arrow. Click and drag the frame of the bitmap until it extends to the 00:00:05 mark on the timeline, and then release the mouse button.

FIGURE 10-22   Select and place a bitmap

5. Go to the Browser window and click Effects. Then click 2D and select the Swirl transition from the top row, on the right in the line of icons, and drag it onto the bitmap object track in the timeline. When the cursor changes indicating that you can place the transition on the timeline, move it to the beginning of the timeline and click to leave it there.

6. Place the cursor over the bitmap object and right-click. Select Object | Video Effects as shown in Figure 10-23.

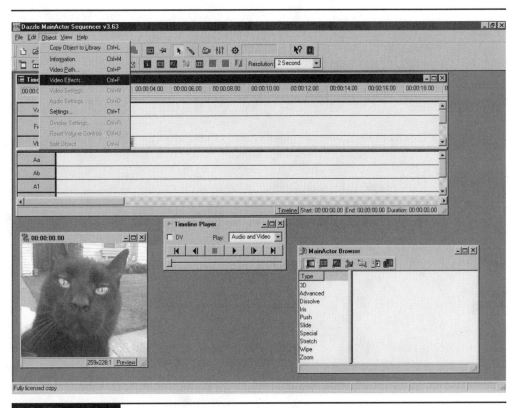

FIGURE 10-23   Object menu

7. In the Video Effect Configuration window, shown in the illustration that follows, select the Swirl effect listed in the Active Video Effects list and click Settings to display the video effects settings window. This window controls which video effects are applied to a particular object. To add an effect to an object, select the effect in the left list box

and click Add Effect. To remove an effect from an object, select the effect in the right list box and then click Remove.

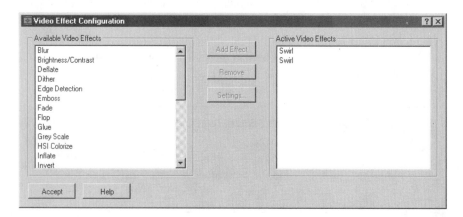

8. To control the settings of an individual effect, select the effect and click Settings to display the video effect settings window, as shown in the illustration that follows. To change the settings, click Accept to return to the Video Effect Configuration window. Click Accept in the configuration window.

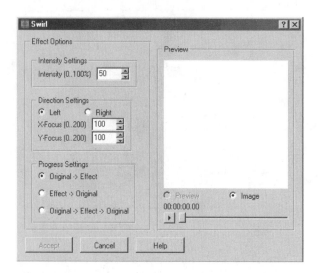

9. You can then add more objects, transitions, effects, and titles to your project. When you have finished, you may save the project or export it to video.

Effects Available in MainActor Sequencer

Most video editing packages above the entry level will contain stock video effects functions. These effects can add interest and creativity to your video project, but should be used thoughtfully and sparingly as it's easy to overdo it.

2D Warp

There are three 2D Warp effects: Glue, Ripple, and Swirl. The Glue effect pulls a clip apart vertically and has the appearance of glue being separated. The Ripple effect ripples a clip as if it were projected on the water's surface when a rock was dropped into it. The Swirl effect swirls the clip around its center, as shown in Figure 10-24.

FIGURE 10-24   2D Warp Swirl effect

10

3D Warp

There are two 3D Warp effects: Deflate and Inflate. The Deflate effect causes the center of the clip to appear to sink like a rubber sheet being pressed down. The Inflate effect causes the clip to appear to expand from the center like a rubber sheet pushed up from below.

Color Adjust

There are four Color Adjust effects:

Brightness/contrast Allows you to adjust the brightness and contrast of a clip.

Greyscale Allows you to change a color clip to a greyscale palette.

HSI Colorize Colorizes a clip using the Hue/Saturation/Intensity model.

Invert Color Inverts the colors of a clip by replacing colors with their complements.

Paint

There are two Paint effects: Oil and Pointillism. The Oil effect applies a brushstroke appearance to a clip so it appears to have been painted with oil paint. The Pointillism effect causes a clip to appear to have been painted using the Impressionist pointillism technique made up of thousands of colored dots.

Perspective

The Perspective effect stretches the clip into the distance, as shown in Figure 10-25.

Special

There are five Special effects:

Light Illuminates a clip as if it were being exposed to a flash of lightning.

Mirror Mirrors a clip at the horizontal midline, as shown in Figure 10-26.

Old Film Causes the clip to resemble a scratchy, old film.

Rustle Moves the clip's pixels around, creating a vibrating appearance.

Zoom Zooms the clip to a specified focal point.

FIGURE 10-25 Perspective effect

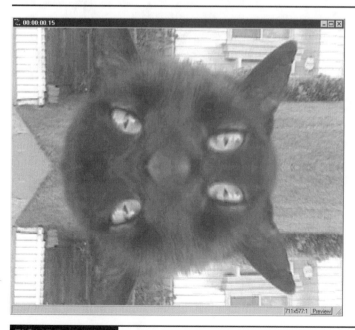

FIGURE 10-26 Mirror effect

Standard

There are five Standard effects:

Blur   Blurs the clip.

Edge Detection   Causes the edges of objects within a clip to appear light, and areas away from edges to appear dark.

Emboss   Gives the clip a raised or embossed look (see Figure 10-27).

Median   Filters noise and specks from the clip.

Sharpen   Enhances the regions where color and intensity changes occur, giving a sharpening effect to the clip.

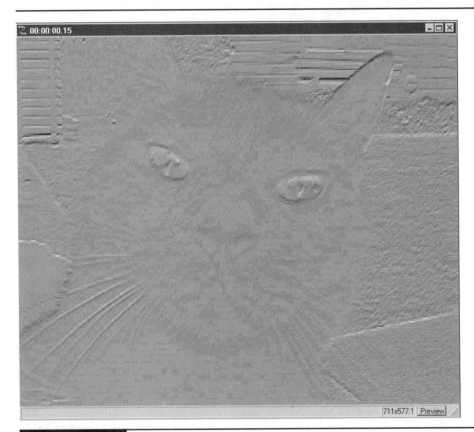

FIGURE 10-27   Standard Emboss effect

Wind

The Wind effect, shown in Figure 10-28, causes a clip to look like it was placed in a high-speed wind tunnel.

FIGURE 10-28   Wind effect

Video Paths

Video paths assign a path created by adjusting a set of control lines and directions in a control window. These controls tell the program how to move the video clip in the final rendered video along the designated video path. These effects are particularly useful in developing opening and closing sequences and title sequences. They can also serve to introduce a product image into a scene when creating how-to and product marketing videos. Most video editing packages above the entry level will contain stock video path functions. Both MainActor and Adobe Premiere include a set. You can also purchase additional paths from third-party software vendors on the Internet for Premiere and other more advanced editing software packages.

Video Paths in MainActor Sequencer

The Video Path Configuration window (shown in the illustration that follows) is displayed by right-clicking an object and selecting Object Video Path from the pop-up menu. To add a path to an object, select the path in the left list box and click Add Path. To remove a path from an object, select the path in the right list box and then click Remove.

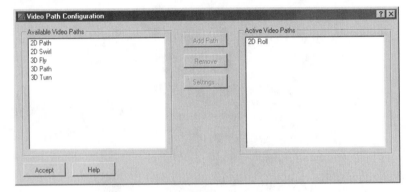

The Settings button displays the video path settings window that controls the settings of a selected individual path. The settings window for 2D Roll is shown here:

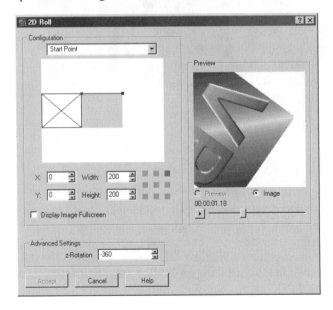

Video paths are heavy-duty effects and should be used sparingly.

2D Video Paths

MainActor has two 2D video paths. The basic 2D path moves a video clip along a specified two-dimensional path. The 2D Roll path allows you to fly in a clip from far away with a spinning motion (see Figure 10-29).

3D Video Paths

MainActor has two 3D video paths. The basic 3D path moves a video clip along a specified three-dimensional path. 3D Fly video path makes a clip appear to fly in from a distance with a spinning and flipping motion.

FIGURE 10-29   2D Roll path

10

Summary

There is certainly much more to learn and know about video special effects and transitions than I have discussed here. New effects are constantly being created. Just remember the point of your video is to communicate ideas, information, and emotion. Special effects are just tools in your kit. They are a means to serve the program, but should never become the point of the program. Effects and transitions are fun, so go ahead and use them, taking care not to let them dominate the video.

Chapter 11

Synchronize Your Video with Sound and Music

How to...

- Use external audio clips
- Do voiceover and other recordings
- Use stock music
- Create custom music tracks with software tools
- Edit audio clips
- Use multiple audio tracks in your video
- Add audio special effects

This chapter covers the capturing, processing, and editing of your video soundtrack. It includes recorded audio and the use of sound effects and music. This chapter focuses on what the professional video world calls *post-production*. Post-production includes adding sound, narration (called *voiceover*), music, sound editing, effects processing, and other aspects of sound after the video has been shot. These features are added as the video is being assembled and edited in your editing software and other support software, such as specialized audio editing and processing packages.

Adding Music to Your Production

Music increases psychological, sociological, emotional, and cognitive effectiveness with an audience. The best example of this occurs within the first ten seconds of a movie. The viewer is able to predict the mood and form expectations of the content based on the first few bars of the overture. The sweet wail of violins warns everyone—get a hanky, tear-jerker coming up. Brass and drums signal adventure and action. Eerie synthesized music announces invaders from outer space. Every composer who scores film and television knows how to create emotional effects and enhance the effectiveness of the experience for the audience. While you may not be up to composing a complex score for your video production, music is a very important element. There are three ways that most amateurs can produce good musical scores for video productions. The first way is to do it yourself if you are a musician or can convince a music friend to help. Failing in-house musical talent, likely ways are to use stock music or an automated music generation software package.

If you are doing a video for home use only, such as your child's first birthday party, you may use popular music from a CD or MP3 file. This would not, of course, be appropriate for commercial distribution.

Using Stock Music

One of the best sources for music when adding soundtracks to your video is the stock music library. These collections of CDs and tapes, with music for every possible situation, are widely available with varying quality. The musical clips may be long compositions used as background tracks for dramatic productions or documentaries, or they may be 10-, 15-, 30-, or 45-second clips used for commercials. The primary reason to use stock music is production cost. Stock music can be very inexpensive. The downsides include difficulty of timing clips to editing, overuse of the tracks in the marketplace, and difficulty finding just the right composition. Almost any style of music can be found, from a country jingle to classical music of the masters.

You may encounter licensing in the form of the "buy out" system, where you simply buy the collection of CDs along with permission to use it in virtually any situation, or it may be a "needle drop" license. The needle drop system requires that a license be paid for each use of a musical clip. There are many inexpensive music libraries designed for use in amateur video productions to be found in the classified and display ads of most amateur video magazines. There is also a huge amount of music and sound of all types available from the Internet. Do a search for the type you need and you will no doubt be overwhelmed with choices.

MP3 files can be easily integrated into your video editing process directly.

If you use performances from a library, copyright and publishing clearances are generally covered. If you license musical performances from record companies, you will probably need to obtain permission separately from the publisher of the musical composition and perhaps the artist who performed the work. If you are recording live performances of a musical work, fees and permissions must be obtained from the artist and publishers. There are very stiff fines and penalties for even minor copyright violations. These permissions should be planned in your production budget and should be obtained early in the process due to time lags in the permission process, or in case of refusals.

NOTE   *The cost of music licensing can be very expensive. Unless you have a significant budget for your project, you should consider MIDI and other royalty-free sources for your music.*

Using Computer-Generated Music

ACID Music 3.0 is a software package from Sonic Foundry, Inc. that will allow you to create royalty-free music for your video music tracks quickly and easily. You can create music in a wide variety of styles by using its multi-genre loop library containing more than 600 loops (genres include: Dance, Hip-Hop, Techno, Industrial, Pop, Rock, Jazz, Ambient, Orchestral, and more). ACID allows you to "paint" your music without having to use complicated MIDI sequencers, drum machines, or musical instruments. You can further enhance each individual track in your musical production with sophisticated and professional sound effects.

ACID, as shown in Figure 11-1, is a self-contained software tool that allows you to create music using a simple "pick, paint, and play"–style interface. You can add MP3 files and songs from other sources to your project (just pay attention to copyrights!). In addition to music created with ACID, you can record music, instruments, or vocals directly into ACID. For those of you with MIDI musical software experience, it allows the use of an unlimited number of loops with automatic tempo and pitch matching. You can preview loops in real time before adding them to your mix, as well as import videos and create music beds. It also allows you to customize your sound with studio effects such as EQ, Reverb, Delay, Chorus, Phaser, and more. You can change the pitch or tempo of an entire project in real time, and mix several tracks into a new one (track bouncing). Finally, you can encode video, insert metadata command markers, launch websites, display captions, and embed URL flips into your audio and video streams.

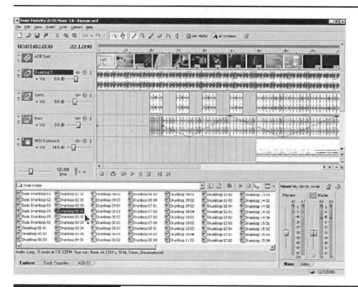

FIGURE 11-1   ACID main screen

ACID has tools that allow you to coordinate the music tracks it creates with a rough cut of your video project. Just finish the video editing and export an MPEG file of your video with the rough audio track, including the voiceover. Using the Video Tracker, shown in Figure 11-2, you can then synchronize the ACID music track production with your cut. When you are finished, you can export the ACID track to a WAV file and insert it into your video editor's timeline. Voila! You will have a royalty-free custom and unique music track just as if you hired a professional composer.

FIGURE 11-2   ACID Video Tracker

   *Select the musical style you want to work with and review a variety of sounds for your sound palette before beginning the ACID scoring process. Sonic Foundry has a wide selection of samples and resources on CD-ROM to work with.*

Recording and Capturing Audio Clips and Elements

You might want to record additional sound effects, music, or voiceover tracks to add to your video production. You will often need to record or capture audio clips or tracks to your computer independently from video clips with integrated audio tracks. This can be done in a number of ways, including:

- Directly to the computer
- Using the audio track of the video camera
- Using a tape recorder and capturing it to the computer separately

If you are recording an audio track on your digital video camcorder, you should use exactly the same process of capturing a DV clip that was described in Chapter 8. You must capture both video and audio tracks and then delete the unwanted video when you import it into the MainActor Sequencer window. In all other ways the process is identical.

If you have recorded to an analog tape recorder, such as a cassette tape machine, you will want to use the Windows Sound Recorder utility or the record function of an audio editor software package such as Sound Forge by Sonic Foundry to record the track to your computer. The following steps show you how to use Sound Forge to edit clips:

1. From your Windows Start menu, choose Programs | Accessories | Entertainment | Sound Recorder. A small control panel will open on your Windows desktop, as shown here:

This utility allows you to record from any analog source that is plugged into the line input or microphone input of your computer's sound card. If you are recording voiceover, you might want to use a microphone plugged directly into the computer.

 *Pay attention to microphone types and quality the same as you would in plugging an external microphone into your video camcorder. Refer to Chapter 5 for guidance.*

2. You will want to check the volume level of your audio mixer from your computer to make sure that the proper input is enabled and the input volume is turned up to an adequate level. These volume control utilities vary with each brand of computer so you should refer to your computer's manual for directions. On my Compaq, it looks like the panel shown here:

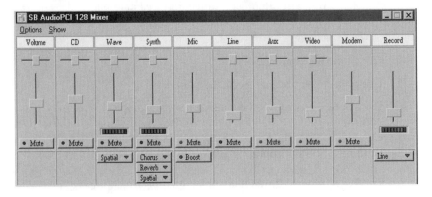

3. When you are ready to record, click the button with the red dot on the right of the control window. This will begin the recording process.

4. Once you are satisfied with it, you will need to save the recorded clip to a file.

Make sure you listen carefully to the recorded clip before accepting it as final. It is easy to overlook subtle background noises such as airplanes flying overhead when you are in the heat of recording.

When recording voiceover it is a good idea to keep a list of your audio takes so you know where you are on a tape or file. Much like a shot list, an audio take list can dramatically speed things up in edit.

Using an Audio Editing Software Package

Today almost all audio editing in professional studio productions is done with computer-based, non-linear editing systems. These software systems are very much like your video editing software packages—in fact the more advanced video editing software packages like Adobe Premiere and Vegas from Sonic Foundry are often used as primary audio editing software. In many cases, you will want a separate audio editing package, like Sound Forge by Sonic Foundry, as shown in Figure 11-3, to process and edit clips of audio and music before inserting them into the video editing system. These packages give more control and allow easier editing and processing of voiceover and music than the tools in many video editors, especially less sophisticated packages such as MainActor. You can purchase excellent sound editors for under $100, including the "light" or limited edition version of Sound Forge. More professional packages can cost several hundreds to thousands of dollars, with many more features included.

11

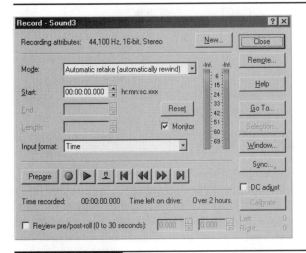

FIGURE 11-3   Sound Forge audio editing screen with Record control panel

Common Editing Operations

The edit operations used most often include Cut, Copy, Paste, Clear, Mix, and Trim/Crop. Most of these make use of the clipboard, which is a temporary storage area that can be used to move data from one window to another. The following list provides a brief description of each operation:

Cut   Deletes a selected portion of data and copies it to the clipboard.

Copy   Copies a selected portion of data to the clipboard.

Paste   Inserts the contents of the clipboard into a data window at the current cursor position.

Clear   Deletes a selected portion of data, but doesn't copy it to the clipboard.

Mix   Mixes the contents of the clipboard with the current data in a window starting at the current cursor position. Mixing allows you to combine two sounds together into one window so you can create complex sound effects.

Trim/Crop   Deletes all data in a window except the selected section. Trimming or cropping allows you to single out a section of data. This is a handy feature since you can keep using the Play button to hear selections until you have just the right amount and then get rid of everything else with the Trim/Crop command in the Edit menu.

Status Formats

When editing sound files, the Ruler, Total Length Status, and Playbar Selection Status can be set to different formats so you can coordinate sound files with other events, or edit to a timing base that you feel most comfortable with. Lengths and positions can be displayed in a variety of formats including Samples, Time (seconds and milliseconds), Frames, SMPTE, and Measures and Beats. This is particularly valuable when coordinating or synchronizing with video tracks.

Editing Stereo Files

When editing stereo files, you have two channels of data on which to work. The upper channel is the left channel and the lower channel is the right channel.

Selecting Data in Stereo Files

When selecting data in stereo files, Sound Forge allows you to select the left channel, right channel, or both channels for playing, editing, and effects processing. When editing a stereo file, the Wave Form display showing the two channels is split into three logical sections for selection with the mouse. The upper section of the Wave Form display is the left channel "hit" section, the lower section is the right channel "hit" section, and the middle is for both channels.

When selecting data with the mouse, which area you are in determines what channel(s) will be selected.

Single Channel Editing

Stereo data files are tied together by their nature and other cosmic forces. In other words, they always play together. This means that there are some edit operations, such as Cut and Paste, that you can't use on a single channel. It would leave one channel shorter or longer than the other. This is usually not a problem in real-world editing situations.

You can copy a selection from a single channel to the clipboard by selecting the data in either the left or right channel and using the Copy command. This will place a mono clip on the clipboard. You can then paste the mono clip to a mono file or to both channels of a stereo file, or you can mix it into a single channel of a stereo file. When mixing mono clipboard data to a stereo file, you will be asked whether you want to mix to a single channel or both channels.

All sound editors work with the same type of graphic interface and the terms and effects are similar. You should have no problem moving from one editor to another or applying the steps described previously to another package.

Editing an Audio Clip with Sound Forge

Let's walk through the process of editing a sound clip with Sound Forge.

1. The first step is to click your Sound Forge icon to open the program. The main screen appears, as shown in Figure 11-4.

2. Select and load a sound file to edit. This is done using the standard Windows file load process (File | Open) and the Open dialog box, shown in Figure 11-5.

3. When you have selected and loaded a sound file you will notice that the wave form of the audio is displayed in a window overlaid on the main screen. The jagged edges of the wave form represent the shape, amplitude, and frequency of the sound that has been recorded. It is a visual indication of the word, sound, or music recorded. This makes it much easier to identify and cut or process all or parts of the audio file. It is very much like the video clips you have seen in the video editing package. You may have a monophonic file with only one wave form or a stereo file with two.

4. Select a section of the wave form by clicking on the begin point you want to select, and while holding the mouse button down, drag to the left until the area you want to select has been highlighted in black. You may select an area of one or both tracks at the same time (see Figure 11-6). This area of the file can be deleted from the file or it can be processed with an effect or filter. Let's add reverb to the selected section.

11

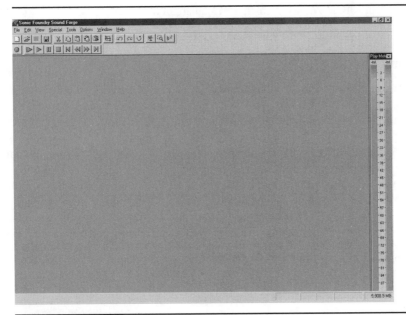

FIGURE 11-4   Sound Forge main screen

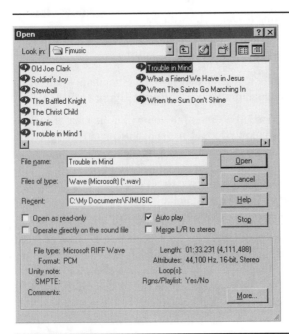

FIGURE 11-5   Sound Forge Open dialog box

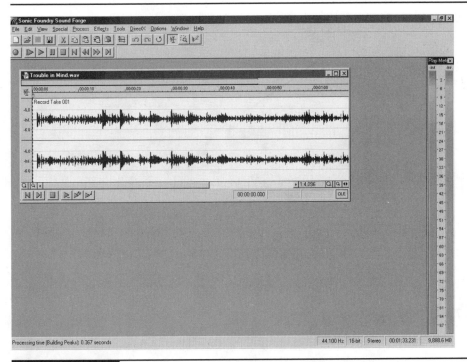

FIGURE 11-6   Wave Form area selected for processing

5. Select Reverb from the Effects menu, shown here:

6. The Reverb control window appears, as shown in Figure 11-7. In this window, you can set up the parameters of reverb you want and then click Preview to hear what it will sound like, or OK to apply. Sound Forge has unlimited Undo if you make a mistake.

11

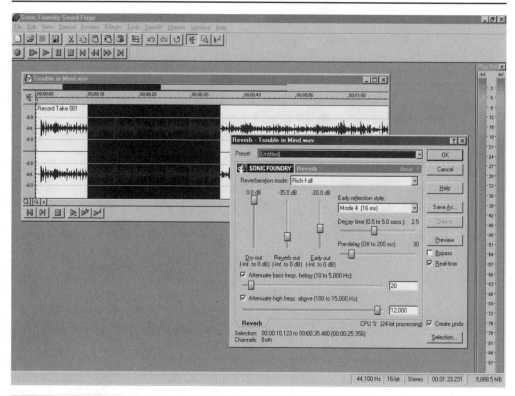

FIGURE 11-7   The Reverb control window

7. You can use the same process to add any effect or filter or you may delete the section with the DELETE key on your keyboard.

8. When you have finished with your editing, you may save the file or use Save As to save it as a new file if you want to retain your original file.

9. You can then insert the edited audio clip into your video timeline.

Processing Audio Effects in MainActor

There are many reasons to process and add special effects to your audio track, such as elimination of noise that has been inadvertently recorded, adjusting the volume and/or quality of the audio track, or creating a special audio effect such as echo. You will need to pay attention to the relationship of the audio track of one clip to the next and to the overall quality and sound level of the final video. You can't have the volume of the audio going from loud to soft and back without distracting the viewer. (Naturally, you may want the sound to get louder on occasion when it is heightening the action, but this should be used very sparingly as a special effect.) The video editing software packages all provide at least a minimal library of sound effects, volume controls, and filters to accomplish these important tasks. This section will discuss some of the possible audio controls available in MainActor. These are similar to those found in Premiere and other packages as well.

Just as with video special effects, audio special effects can overwhelm a production. Use effects sparingly and with careful thought. Reverb and echo are great in a small dose, but obscure voices when used heavily.

When using MainActor and similar software packages, control windows are provided that allow for the selection and control of the parameters of audio effects such as the Band Pass filter or the Echo effect. The window's appearance will vary depending on the particular effect, but the settings on the right such as the mark-in/mark-out settings and the preview section are common to all versions of this window.

The Audio Effect Configuration window, shown in Figure 11-8, is opened by clicking the Audio Effects icon on the toolbar or by clicking the audio track and selecting Effects from the pop-up menu. You can apply an effect to the audio track by highlighting it in the left window and clicking Add Effect. This will move the effect to the right window.

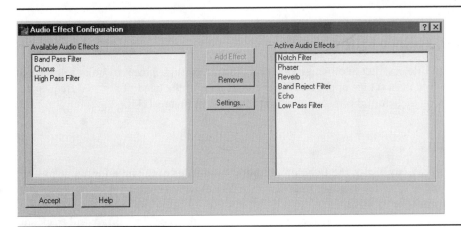

FIGURE 11-8   Audio Effect Configuration window

The Audio Effect Configuration window allows you to apply effects to an audio object. The following sections describe the effects that are available.

Band Pass Filter

The Band Pass filter removes all frequencies from an audio track or segment except those that lie within a specified range (band). You can set the upper and lower frequencies in the Band Pass Filter control window shown in Figure 11-9. When choosing a bandwidth, keep in mind that the higher the bandwidth the more the surrounding frequency range will also be affected.

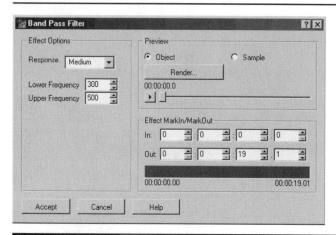

FIGURE 11-9   Band Pass Filter control window

Band Reject Filter

The Band Reject filter allows all frequencies to pass through except those in a specified range (band). You must set the upper and lower frequencies of the band of rejected frequencies in the settings on the left of the control window, as shown in Figure 11-10.

Chorus

This effect is similar to the Echo effect, but it modulates each wave form's echo with a configurable wave form (none, sine, square, triangle) with a configurable speed and amplitude. This isolates the echoes.

The Chorus effect is so named because it makes the recording of a vocal track sound as if it was sung by two or more people in a chorus. This is achieved by adding a single delayed signal (echo) to the original input. However, the delay of this echo is varied continuously between a *minimum delay* and *maximum delay* at a certain *rate*. Typically, the delay is varied between 40 milliseconds and 60 milliseconds at a rate of about 0.25 Hz.

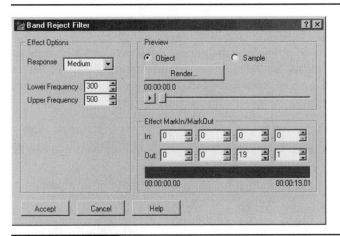

FIGURE 11-10   Band Reject Filter control window

Some of the configurations shown in the Chorus control window (see Figure 11-11) include:

Modulation type   The type of the modulating wave form.

Modulation speed   The frequency in Hz of the modulating wave form.

Modulation depth   The amplitude of the modulating wave form.

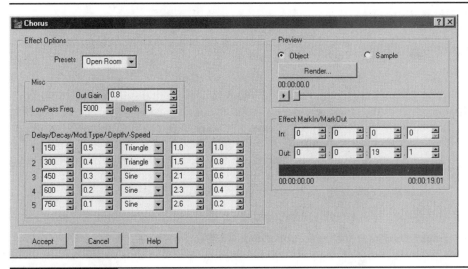

FIGURE 11-11   Chorus control window

11

Echo

Echo is produced by adding a time-delayed signal to the output of a sound processing device or filter (see Figure 11-12). This produces a single echo. Multiple echoes are achieved by feeding the output of the echo unit back into its input through an attenuator. The attenuator determines the *decay* of the echoes, which is how quickly each echo dies out. This arrangement of echo is called a *comb filter*.

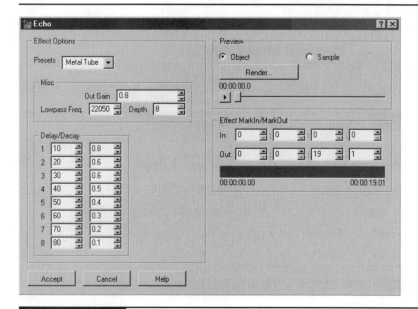

FIGURE 11-12   Echo control window

Echo greatly improves the sound of a distorted lead guitar solo, because it improves the sustain effect of the music and gives an overall smoother sound. Very short echoes (5 to 15 milliseconds) with a low decay value added to a voice track can make the voice sound "metallic" or robot-like. This was a popular way of creating robotic-voice effects in old movies.

This effect adds "<amount> of <delayed>" and "<decayed>" echoes on the input stream. The "< >" indicate what the variable is in the process that can be controlled in the control window.

Some of the configurations available in the Echo control window include:

Out Gain   Indicates the global percentage of all echoes present in the output stream.

Depth   Indicates the number of echoes.

Delay   Indicates the delay (in milliseconds) of the echo.

Decay   Indicates percentage of the input stream volume this echo should have.

High Pass and Low Pass Filters

The High Pass filter allows all frequencies above a given frequency, known as the *cutoff frequency*, to pass through and suppresses all others. This is used much like the treble control on your stereo system to reduce or emphasize the higher frequencies of your audio track. The control window for this filter is shown in Figure 11-13.

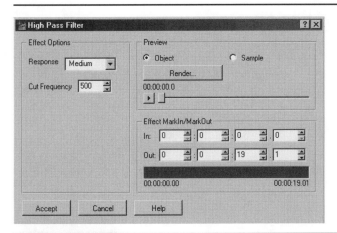

FIGURE 11-13   High Pass Filter control window

The Low Pass filter allows all frequencies below the cutoff frequency to pass through and suppresses all others. This is used much like the bass control on your stereo system to reduce or emphasize the lower frequencies of your audio track. The control window for this filter is shown in Figure 11-14.

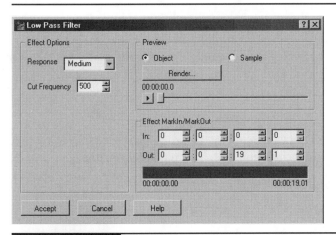

FIGURE 11-14   Low Pass Filter control window

Notch Filter

The Notch filter allows the user to specify a specific frequency to sharply reduce or emphasize a sound. This might be used to reduce a 60 Hz hum picked up in the audio signal from an electric motor or other source. It can also be used to emphasize a voice over a musical background by emphasizing the voice frequencies. The control window for this filter is shown in Figure 11-15.

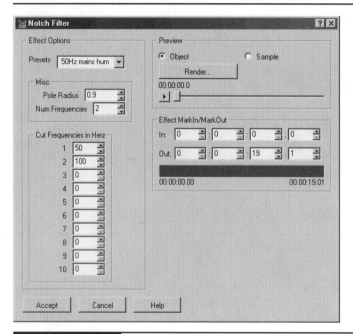

FIGURE 11-15   Notch Filter control window

 *Use the Notch filter to reduce AC line hum from your audio track. This can be a boon if you are working with material that is difficult or impossible to rerecord.*

Phaser

A Phaser combines the Reverb and Chorus effects. If two identical but out of phase signals are added together, they will cancel each other out. If, however, they are partially out of phase, partial cancellations and partial enhancements occur. This leads to the phasing effect. Other strange effects can be achieved with variations of Reverb, Echo, and Chorus. The control window for this filter is shown in Figure 11-16.

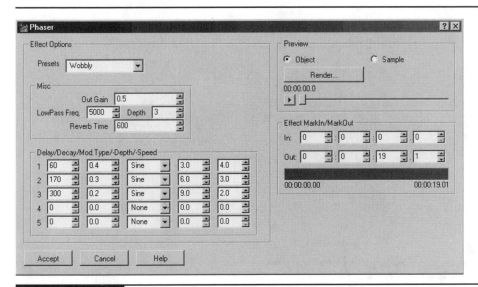

| FIGURE 11-16 | Phaser control window |

Reverb

Reverb is used to simulate the acoustical effect of rooms and enclosed buildings. In a room, for instance, sound is reflected off the walls, the ceiling, and the floor. The sound heard at any given time is the sum of the sound from the source, as well as the reflected sound. An impulse (such a hand clap) will decay exponentially. The *reverberation time* is defined as the time taken for an impulse to decrease by 60 db (decibels) of its original magnitude.

This effect is similar to the Echo effect, but it feeds the output stream back onto the input stream with a given "<reverb delay = reverb time>" which creates a richer sound (multiplies the echoes). You will achieve the best results when the reverb delay is double the largest echo delay, and the echoes are evenly distributed.

One of the options available is Reverb Time (see Figure 11-17), which indicates the number of milliseconds the feedback of the output stream is delayed until it is added onto the sample stream again.

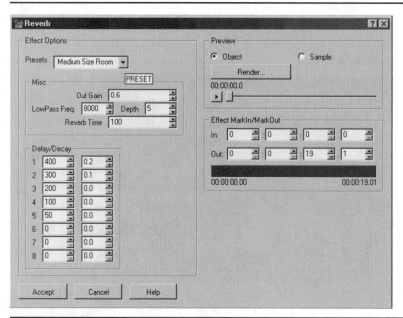

FIGURE 11-17   Reverb control window

Inserting and Deleting Audio Tracks in MainActor

You can insert one or more additional audio tracks into your video production in MainActor and most other editing software packages. This will allow you to add music or other effects to the sound you originally recorded. You can also remove the audio track originally recorded and replace it with another or leave it out all together. This is useful when you intend to supply the audio with music or with voiceover recorded later.

Adding and deleting an audio track is essentially the same as adding or deleting a video track.

1. Right-click the additional audio track you want to insert (for example, Ab) and select Insert Multimedia, as shown in Figure 11-18. This will open the Import Dialog and allow you to select the file you wish to insert in the track, as shown in the illustration on the following page.

2. Click OK if you wish to accept the defaults or select different options as desired. This will open the Select File(s) dialog box.

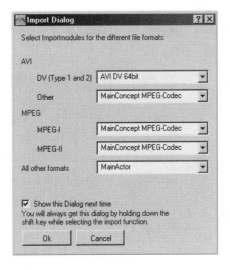

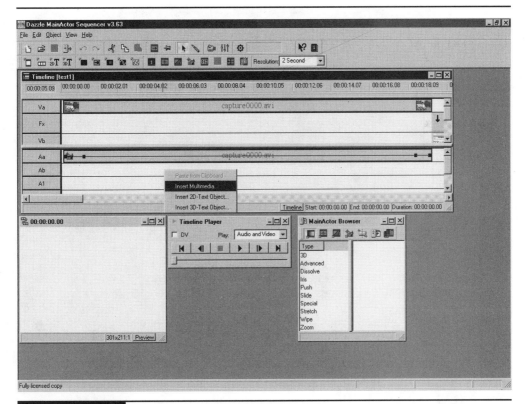

FIGURE 11-18   Insert Multimedia command

3. Select an additional or replacement audio track to insert as shown in the following illustration. Click Open to insert the track. The new track is now shown in audio track Ab, as shown in Figure 11-19.

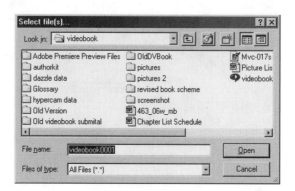

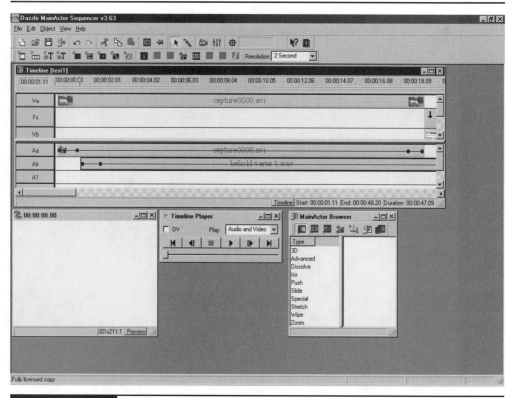

FIGURE 11-19   The new track in audio track Ab

4. You can delete the track by right-clicking and selecting Delete, as shown in Figure 11-20.

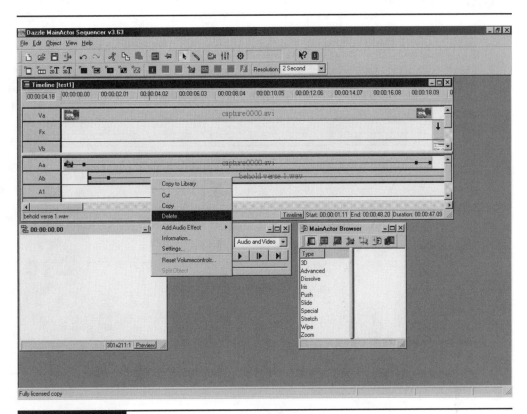

FIGURE 11-20   Deleting an unwanted audio track

Trimming Audio Clips

You can trim your audio clips in an external sound editor such as Sound Forge or you can insert your audio clips into your video editor and trim them there. In MainActor it works like this:

1. Select the Edit razor tool from the MainActor toolbar (see Figure 11-21). It will convert the arrow of your cursor into an icon that resembles a razor knife.

2. Select the beginning and end points of your cut by clicking on the audio track. You can then delete unwanted sections that you have selected with the razor tool.

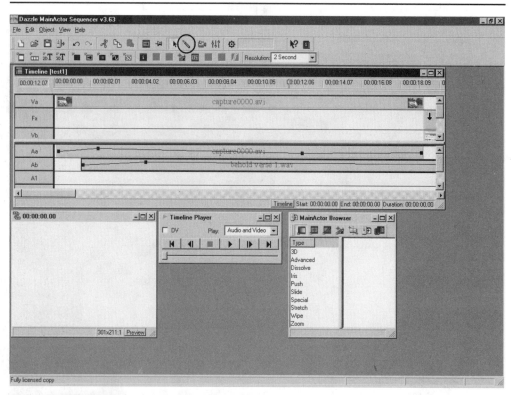

FIGURE 11-21   MainActor trim process

Mixing Soundtracks and Balancing Dynamics

One of the main things you will need to adjust in your audio tracks will be the volume or levels. This will determine how loud or soft each track is when played back and will govern the mixture of the separate tracks or the primary track if you only have one. The relationship between the level of one track and the other tracks in the video is also important. You should not allow dramatic differences between them unless of course it is to make a deliberate point. It is very easy in most editing software to adjust the volume levels of the tracks.

MainActor and Premiere both show a thin line with small black squares, called *nodes*, along the lines overlaid on the audio track object (see Figure 11-22). These lines indicate the relative level of the volume, similar to a sliding volume control on your stereo. If you use your mouse to move the line up, the volume will increase, and moving it down will decrease it. You can see if the volume is the same at the beginning and end of the tracks and adjust them to match the adjoining tracks' levels. You can also grab one of the nodes and drag it on a slant, thereby creating a gradual increase or decrease in sound level, called a *fade*. You can fade in or out by increasing or decreasing the volume level.

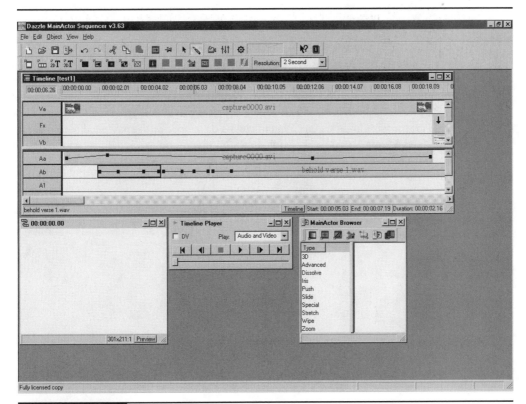

FIGURE 11-22   Audio track level setting lines and nodes

Make sure that the levels at the beginnings and ends of adjoining clips are matched in volume to avoid clumsy and abrupt sound changes.

Synchronizing with Recorded Video Sound

You might need to synchronize the sound or music from one track with another. This can be done by simply dragging the track to the left or right and previewing the result until you get the two or more tracks properly synchronized. If you are synchronizing with spoken voice on the video clip, you can review the lip synch in the video window and adjust it until the two are properly synchronized. This can take a while, so be patient and trust that the results will be worth the attention to detail.

Summary

Sound can add tremendously to the effectiveness of a video project. Music can signal the emotional mood and content of the video. Sound effects can be a fun addition to a soundtrack, and of course, voiceover is an essential component of documentaries and can also be fun in less serious video projects. Music and sound effects are available from many different sources, including inexpensive libraries on CD, and can be developed with automated software packages. Capturing and editing sound works very similarly to your video editing software package and a dedicated audio editing package can be an asset in your video making toolbox.

How to...

- Design titles for your video projects
- Create titles with a paint program
- Make a scrolling 2D text title
- Create a 3D animated title

There are many approaches to creating titles for video and film. Entire industries in Hollywood are dedicated to creating the most original and stunning titles for movies and television productions. The best way to get ideas is to pay close attention to the way these professionals use beginning and end titles in their productions. It is also a good idea to look at documentary and training videos as well as other projects that may relate to your own video projects and see what kind of approaches they have taken. This chapter will cover the design and creation of a variety of titles and title types, just scratching the surface of the possibilities.

Designing Effective Titles

While it is important to do a good job on your titles, it isn't necessary to go to the creative frontiers. The key elements include:

Clarity   Give all the necessary information and credits in your titles. On the other hand, don't make the titles overwhelm the production itself. It is usually better to include the thanks to Grandma and cousin Bob at the end rather than the beginning.

Appropriateness   If you are doing a documentary video on the history of your church, it will probably not be suitable to use animated dogs carrying title cards around their necks for titles. On the other hand, for a dog-lover's birthday party video (or for the dog's birthday party), they could add quite a dash of panache.

Easy-to-read fonts   Use a clean font with a carefully selected color. Make a sample and look at it on a TV set to make sure it is easy to read and the appropriate size and layout.

Pay attention to over-scan   Video signals, when played back on a TV set, extend beyond all four edges of the screen. This phenomenon is called over-scan. Make sure that you place your titles in the active area of the screen and don't extend into the over-scan areas. You can check this out by making a sample. In some editing programs, there are title-safe generators to make sure titles don't get cut off. The general rule is titles should not exceed 80 percent of the screen area.

You should try to have fun with your titles and not be intimidated by the prospect of designing and creating them. If you keep them clean and simple, it is hard to go wrong. I will cover a number of title creation techniques, primarily using MainActor. There are many other software systems designed specifically for creating titles for video. If you have access to one of these you might add it to your video arsenal.

Examples of title creation software packages you might consider acquiring include Pinnacle Systems' TitleDeko and Inscriber TitleExpress. These packages offer more features and controls than the title functions in most basic editing packages and are useful if you are creating more sophisticated titles.

Creating Opening Titles

The first thing your audience will see is your opening title sequence. This will set the tone and create expectations for the balance of your video. It should be as artful as possible and should convey the style and tone of the project. Artful does not necessarily mean fancy. In fact, nothing says amateur like over-produced titles. Create titles that are in keeping with the style of your production, and emphasize simple, clear, and exceedingly well made, rather than too busy and poorly executed. The selection of music or sound underneath the opening title is also an important issue and should carefully match the graphics.

There are countless opening title possibilities, from simple title slides over black to dazzling animated text flying onto the screen. Much of what you can imagine is within the realm of possibility with the use of computer video editing and animation software. Basic title and 3D text animation is even included in inexpensive video editing software like MainActor, and their results, with ease of use, can be spectacular and professional. In a professional production, opening titles typically include:

- The producer
- The director
- The production company
- The distribution company
- The title of the video
- The stars

In a simpler video project, you will probably not need all of these categories—until you start making professional videos, of course. In a project like your child's first birthday party, you may find it fun to include 3D animated titles using some of the following:

- A Bob Smith Production
- A Bob Smith Video
- Bob Smith Productions
- Distributed by Bob Smith Films
- The Adventures of Bob Junior's First Birthday

12

Of course, you can simply put a title at the beginning describing the event, but titles can be fun and add a lot to even a basic production. Make it fun if the video is fun, or formal if it is formal. An example is shown in Figure 12-1.

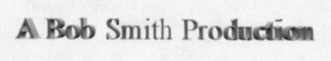

FIGURE 12-1   Basic title using video editor

Another type of opening title, and one that is common in short video projects, uses a series of bitmap graphic files created in a paint package and faded in from one slide to the next, as shown in Figure 12-2.

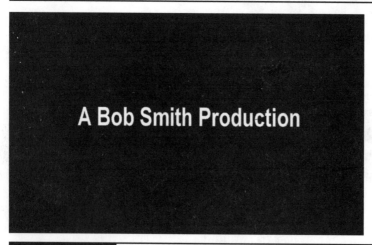

FIGURE 12-2   Basic title using paint package

A similar effect can be achieved by overlaying fixed text over a photograph or textured background, as shown in Figure 12-3.

FIGURE 12-3   Overlaid text title

Pay attention to the font you select and the color you select for the overlay. When you are overlaying a fixed image, you have control over the readability and contrast of the text with the background image. When you are overlaying text over a video clip that is changing significantly, the readability of the title can become problematic as the color or contrast of the background changes. Figure 12-4 shows an alternative font that is less appropriate than the original to the style of the video subject. See the difference?

Figure 12-5 is an example of a text color that doesn't give sufficient contrast with the background. The results render the text unreadable. Figure 12-6 shows a font that might be appropriate style-wise, but its spindly nature renders it unreadable.

Making good titles involves many elements, and making really good titles can be difficult. This should not discourage you from experimenting with the available tools in making titles for your video productions. Very simple titles can be effective, such as the use of text over a black background style, used so well by documentary filmmaker Ken Burns.

12

FIGURE 12-4   Alternative font title

FIGURE 12-5   Inappropriate color use in title

FIGURE 12-6   Weak font in title

 *Be sure to select a font that communicates the style of your production while providing readability.*

Creating Credits

Certainly giving credit is a prime motivator and a reward for work well done. This is never truer than for amateur video productions where the only pay is often a line in the credits and bragging rights. Be generous with the listing and don't make the title crawl so fast that no one can read his or her name. The credits, in most productions, come at the end of the tape, though there is no video law that demands this. Often, sponsors receive a credit in the opening sequence. If you are shooting a documentary at a local museum, or have received particular help from an organization or local business, you might consider this appropriate.

A well-designed arrangement for a credit slide with a minimum of credit lines might look like the one shown in Figure 12-7. An alternative arrangement with more detailed credit lines is shown in Figure 12-8. In both cases I have placed the credit text in the middle of the screen. This will solve the problem of missing text because of over-scan.

FIGURE 12-7   Basic credit slide

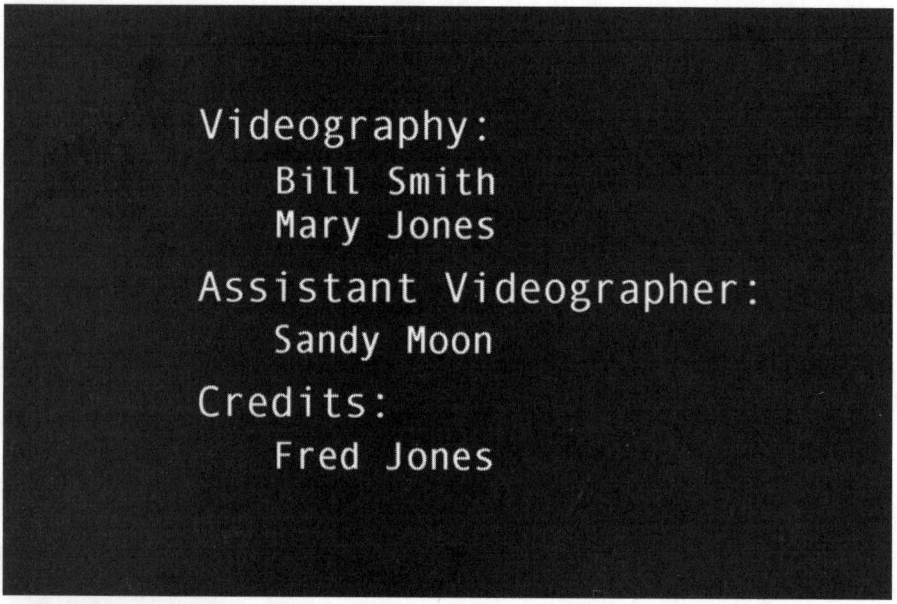

FIGURE 12-8   More detailed credit slide

 *Remember to look at credits on a variety of video productions for ideas. I spent a half day preparing for writing this chapter by looking at short films and video on the Sundance television channel just to get an overview of titles and credits.*

Copyrights and Copyright Notices

You will want to include a copyright notice on the package or label of your videotape or CD-ROM as well as within the credits and titles. Typically this is done on the last section of credits or on a separate title slide at the end of the production.

 *A basic copyright notice will consist of: Copyright © 2001 (use the actual year you create your project), Your Name, and All Rights Reserved. This will cover most situations.*

Using the Windows Paint Program

Let's create a set of simple titles using the basic Windows Paint software utility. This is the easiest and most basic way to create titles.

1. From your Windows Start menu, select Programs | Accessories | Paint. The Paint program will open. Figure 12-9 shows the Paint main screen.

2. Select Image | Attributes from the menu bar. The Attributes dialog box appears, as shown in the illustration that follows. Type in the image size 352 pixels by 240 pixels. This will create a white background for your image the size of an MPEG video frame.

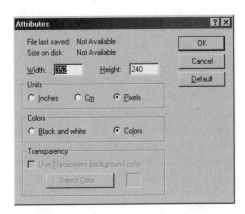

12

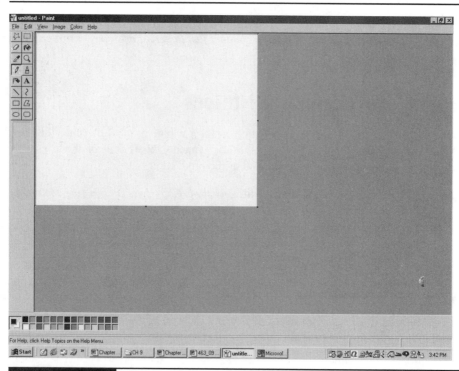

FIGURE 12-9   Paint main screen

3. Click the A toolbar icon. This icon enables you to insert text into your picture. A small crosshair will appear. You must select an area to contain your text. This is done by clicking the screen a bit to the left and above the top of the image and then dragging the mouse to the right and down until you create a dashed square about 1/4 of the size of the screen (see Figure 12-10). Don't worry if it is too small or large or in the wrong place when you add the text. This can be adjusted later.

4. Once your text square is created, the cursor turns into a large I beam when moved into the text square. This indicates that text can be placed if you click the mouse. When you click, a blinking cursor will appear in the upper left of the text square. You may now enter your text. Type **DIGITAL VIDEO** in all caps. Right-click within the text box and select Text Toolbar from the bottom of the menu box. This will open the Fonts dialog box, which allows you to select the font, size, and other attributes (see Figure 12-11). I have set the attributes for Arial Black and the font size at 24. If you prefer a different text or background color for your title, you may adjust the attributes of your image accordingly. Refer to Paint's Help for specific information about using the utility.

TIP   *Preview for readability before completing your titles.*

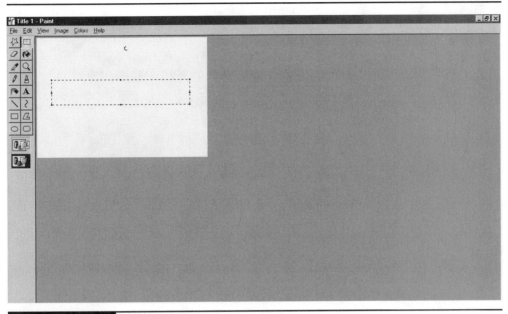

FIGURE 12-10   Inserting a text square in your image

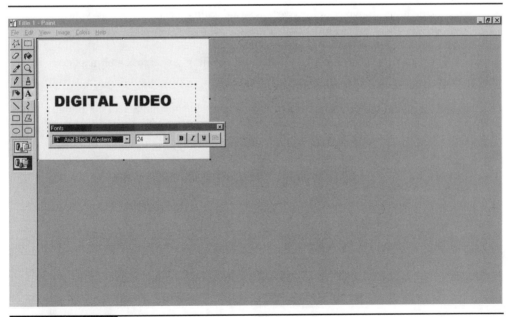

FIGURE 12-11   Fonts dialog box

5. Save your file in a format of your selection in your working directory for your project. I selected the 24-bit bitmap format (as shown in the illustration that follows), but any of the selections will work.

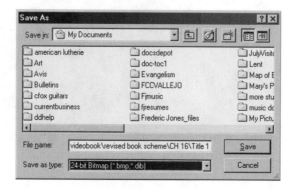

6. Create as many titles as you need for your project using the same procedure. The level of detail is up to you.

Scrolling Text Across a Bitmap

One of the most simple but effective ways of creating a title is to scroll your title text over a bitmap image or overlay it on a clip of video. The first title we will create here will scroll text over a bitmap background.

The step-by-step instructions for creating scrolling text across a bitmap are:

1. Start Dazzle MainActor Sequencer.

2. Right-click the Timeline and select Insert Multimedia. Select the File menu item from the pop-up menu and the Select File(s) dialog box will open, as shown in the illustration that follows. Select an image or bitmap file.

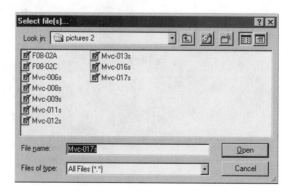

3. When the bitmap file opens, the Dazzle MainActor Sequencer cursor will change to indicate where the bitmap file can be placed. Move the cursor to the beginning of the Va track and click to place the bitmap file on the Timeline (see Figure 12-12).

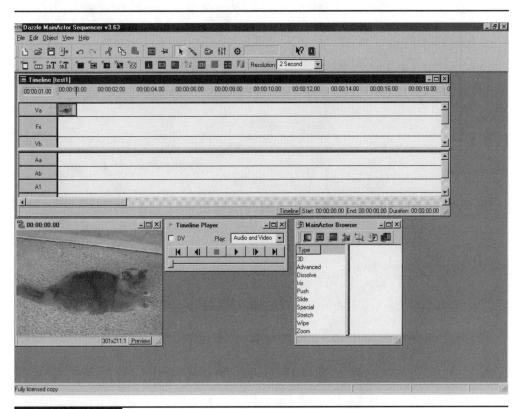

FIGURE 12-12   Inserting a bitmap in the MainActor Timeline

4. Move the cursor over the right side of the bitmap object. The cursor will change to an arrow. This indicates that you can resize the bitmap object. Click and hold the mouse button, then drag the frame of the bitmap object until it extends to the 00:00:05:00 mark on the Timeline. Release the mouse button. This will extend the play time of the bitmap to 5 seconds (see Figure 12-13).

TIP   *Make sure the scroll time is long enough so the titles are easy to read, but not so long that it becomes boring.*

12

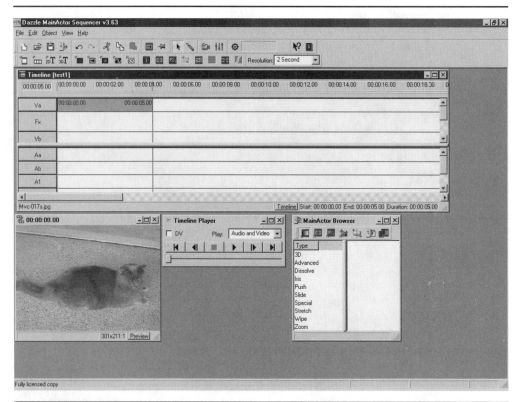

FIGURE 12-13   Adjusting the play time of the bitmap

5. Right-click the Timeline and select Insert 2D Text Object. The Text Object Configuration window will appear, as shown in Figure 12-14.

6. Make sure that the Enable Text Motion box is checked, then type a line of text into the box. Decide how you would like the text to scroll, and then select the starting and ending points for the text. Click Accept Text to finalize your choice.

7. Once this step has been completed, the cursor will change, indicating that the text object can be dropped onto the Timeline. Drop the text object in the V1 overlay track by clicking on the selected spot, as shown in Figure 12-15.

8. To create your video file, select File | Export. When the Export dialog box appears, change the settings appropriately, then click Save. A File dialog box will appear allowing you to set the name of your video file. Click Save in the File dialog box to display the Export Progress window, which will show the video as it is being rendered to the file.

9. You can then add this title element to your project Timeline.

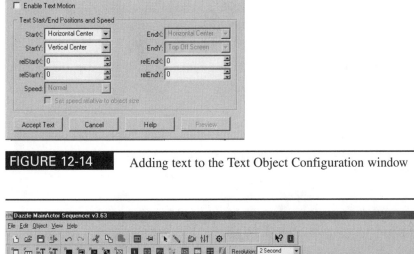

FIGURE 12-14   Adding text to the Text Object Configuration window

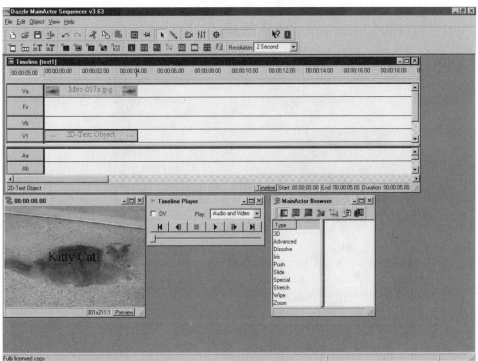

12

FIGURE 12-15   Inserting text into the Timeline

Using 3D Animated Text

Scrolling two-dimensional text is a good basic title process, particularly for end credits and other text-rich applications. When you need to do opening titles or more inspiring section titles, you might consider creating a title with 3D animated text. This can be done easily within MainActor or you can use a dedicated 3D animation software package such as Caligari TrueSpace.

The MainActor Sequencer has a built-in function that allows you to create three-dimensional animated text you can insert into your video project. There are a wide variety of title types that can be created with 3D text. The best way to conquer them is through experimentation.

The process to create a 3D animated text clip is as follows:

1. Open the MainActor Sequencer.

2. Select the 3D icon from the horizontal toolbar. This will open the 3D Text control window, shown in Figure 12-16.

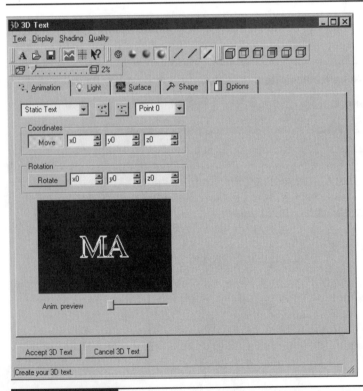

FIGURE 12-16   MainActor 3D Text control window

3. The next step is to import or type in the text you want to animate. Selecting the Text menu option from the Text control window menu bar will allow you to type and import the text. You will need to select New to open the window in order to type and format your text. The default message is "MainActor!" Let's change that to "DigitalVideo!"

4. You can change the attributes of your text by selecting the font, size, alignment, and so on from the option buttons and menus on the Font Dialog (see Figure 12-17). Click Accept when you have finished.

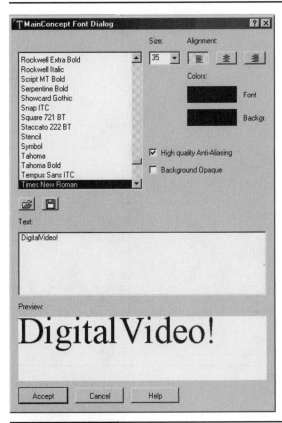

FIGURE 12-17   Font Dialog

There are five tabs on the 3D Text control window that allow you to select the properties of your animation:

Animation   The Animation tab (shown in Figure 12-18) allows you to select the type of path your text object will follow as it moves. There are four choices: Static, Scroll Horizontal, Scroll Vertical, and Flying Into Space. If you click Play in the Preview window you can preview each of these options and select the one you want. To change the path of the animation, you can drag the path line within the Anim. Preview window if the path type you have chosen gives you this choice. You can also click the path icons to the right of your path choice and add nodes to the path to make it more complex. There are path point and rotation coordinates that can be customized. The best way to understand these is to change the setting and see what happens. The number of possible patterns is immense.

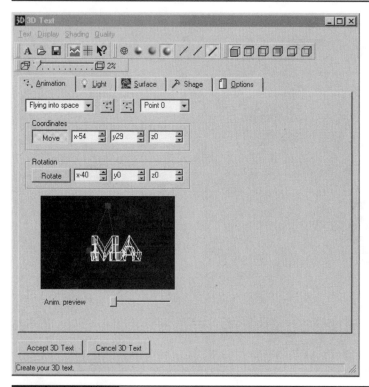

FIGURE 12-18   Animation tab

The paths and the speed of the animation have enormous impact on the readability of your titles. Make sure that you have previewed and adjusted these attributes with particular care.

Light   The Light tab allows the control of the lighting of your text object (see Figure 12-19). There are several preset lighting options available from a pull-down menu. To the right of the menu are two buttons that allow you to increase or decrease the light intensity. You can also experiment with the other parameters to discover new effects such as diffuse lighting or rotating light sources. The choices can be fun, but remember the design points we've discussed, particularly readability.

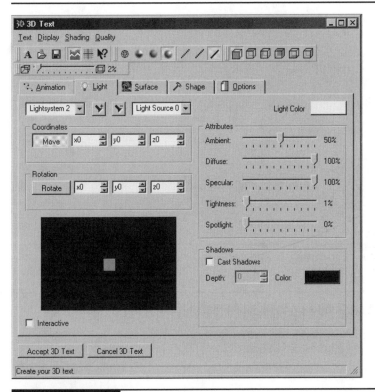

FIGURE 12-19   Light tab

Surface   The Surface tab allows the selection of the surface of your text object (see Figure 12-20). The surface of your text object can be a color, a texture, or a bitmap. You can select these choices and preview them under this tab.

Shape   The Shape tab (shown in Figure 12-21) allows you to select the depth or thickness of the text's three dimensions.

Options   The Options tab (shown in Figure 12-22) allows the selection of the background color and the color of the object's wireframe outline.

12

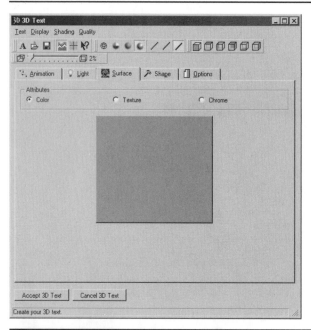

FIGURE 12-20   Surface tab

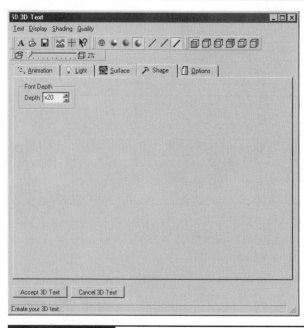

FIGURE 12-21   Shape tab

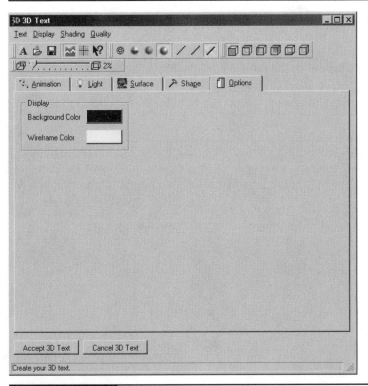

FIGURE 12-22   Options tab

There are a number of other parameters that can be set using the buttons across the top of the 3D Text control window. They include modeling, object orientation (or camera view), the quality of the rendering, and more. Again, the best way to learn these is to experiment.

The final step is to click Accept 3D Text. This will close the 3D Text control window and change the shape of the cursor in your Sequencer window. You can then place the 3D text animation you have created onto one of the Timelines (Va or Vb), as shown in Figure 12-23. Change the duration of the animation by grabbing it on the right edge and dragging it to the time length you prefer.

You can create a variety of title elements and string them together into a complete title sequence. Make sure that the beginning and end points and the transitions between one clip and the next work smoothly and integrate with the sound and music tracks. A sloppy title sets the wrong tone for your video production.

12

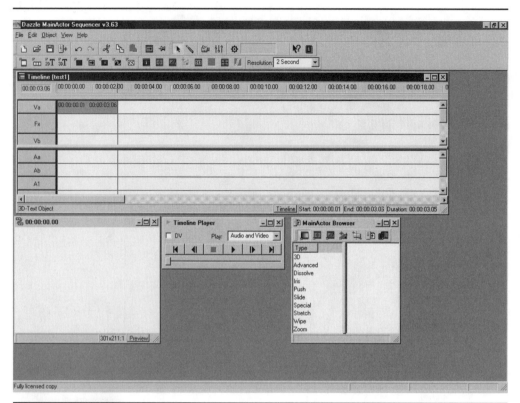

FIGURE 12-23   Placement of 3D text object

Summary

There are many options for 3D text and other 3D animation. Integrated tools such as the 3D Text tool in MainActor provide a great deal of power for video production and are simple to use. The dedicated 3D software animation packages can be daunting and require weeks or even months of practice to become proficient. Remember, major television programs such as *Babylon Five* were animated with personal computer-based 3D animation packages. The unbelievable is possible if you are willing to climb the learning curve.

Chapter 13

Add the Finishing Touches to Your Video

How to...

- Duplicate your tapes for distribution
- Create labels and packaging for your finished videos
- Create CD-ROM labels and packaging
- Properly affix your copyright notices

The difference between an amateur production and a professional one can often come down to the finishing touches. You have done a bang-up job on your video production and editing and the subject is of great interest to your potential audience. The next step will be getting them to pick up the tape or CD and play it. This often comes down to the packaging. We will discuss packaging for both tapes and CDs and a little more on the subject of copyright and copyright notices. You will need these on the physical packages as well as on the tape itself.

Duplicating Your Tapes for Distribution

If you are distributing your video on tape, there are a few things you need to know in order to create a good quality master and copies. The cheapest way to duplicate a few copies of your tape for sharing with others is to connect your digital camcorder with your standard VHS videocassette recorder and make copies in real time. A second process is to make a master tape and have the duplication done by a professional duplication service. These are not much more expensive than do-it-yourself, and they can provide dozens, hundreds, or thousands of copies while you sleep. Let's cover the do-it-yourself issues first.

Connecting Your Camcorder to Your VCR

The best solution is to use a stereo video AV cable designed to connect two VCRs for dubbing. These can be purchased at most video stores, Radio Shack, or other places that sell audio visual cables. The cables consist of three conductors with RCA male connectors on both ends. The color codes are generally yellow for video, white for the left audio channel, and red for the right audio channel. Connect these to the analog RCA video and audio output sockets on your camcorder and to the video audio input sockets on your VCR. You will need to check your VCR instruction manual for the specific details of recording from the selected input connectors. Once the two devices are connected, you should start playing the camcorder a few seconds before beginning the recording process on the VCR.

TIP   *Rather than using standard commercial videotapes that might leave you with a lot of blank tape at the end, it is cheaper and more professional to purchase a stash of tapes of the appropriate length for your project. You can purchase VHS copy cassettes with a variety of tape lengths (5 minutes, 10 minutes, 15 minutes, 30 minutes, and so on). You can get these from professional video supply stores, camera stores, and similar retailers on the Internet. You can also purchase cardboard or plastic boxes to place the finished tape into for distribution.*

Using a Professional Duplicator

A professional duplication service can be a boon for large batches of tape copies. There are, however, some rules you will need to follow. Many duplication services can copy from miniDV tapes. In this case you will only need to take the finished master tape to the service and they will do the rest. It is best for them to make the VHS copies directly from the digital master so that the quality is highest. When you make analog copies from analog copies, each successive generation of copies loses quality. When you make digital-to-digital copies, the copies are exactly the same quality as the original. If you can't find a service that can make copies from your miniDV master, you will have to use the procedure above to make an analog VHS master for them to duplicate from. Check with the service for their particular needs before proceeding.

Make a second or third copy of your tape from your computer so that you will have a backup of the project. Remember, the digital tape copy from your computer is exactly the same quality as the original on the hard drive. There is no "generational" loss or reduction in quality caused by making digital-to-digital copies.

Creating Labels for Your Videotapes

If you have used a duplication house, they will probably be able to provide you with tapes, boxes, and labels to order. If you would rather print your own labels using your computer, it is a very simple process that can be done from Microsoft Word or other word processing package. Many of the label companies, such as Avery, have software utilities for making labels. The following steps use Avery labels and Microsoft Word:

1. Open Word.

2. From the menu bar select Tools | Envelopes And Labels. This will open the Envelopes And Labels dialog box shown here:

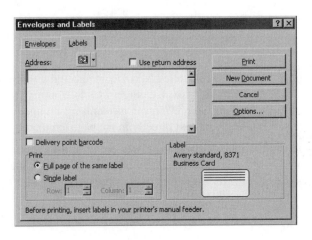

13

3. Click the label icon in the lower-right corner. The Label Options dialog box will open, as shown in the following illustration. Select Avery 5199-F for the video face label template (or other appropriate label product).

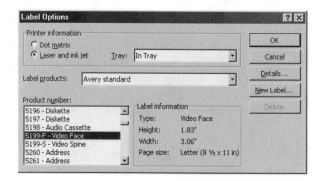

4. This will load your chosen label type in the Envelopes And Labels dialog box. Enter your text data into the Address window, as shown in the illustration that follows. (Even though you are making a video label, not an address label, the procedure is the same for the most part.)

5. When your text is entered, click New Document and a full page of the label templates will be created with your text (see Figure 13-1). You may then format the text's font, color, orientation, and so on. You can also create an empty template by skipping step 4 above and entering your text directly in the page template.

6. When you have created your label document, print a test copy and compare it with the layout on your actual labels for accuracy of your printer. You may need to make adjustments to the text layout or to your printer settings for best results.

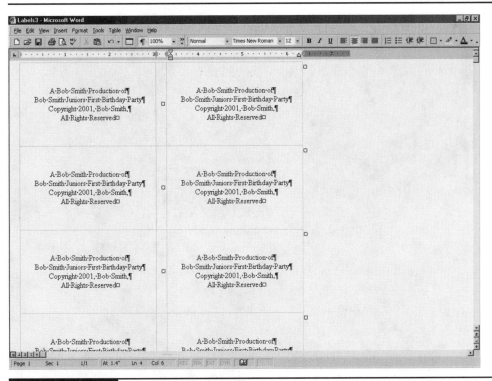

FIGURE 13-1 Format your label for printing

7. When you are satisfied, load your printer according to its instruction manual and the instructions on your label package and print.

Follow these same instructions to print the spine labels for your tapes as well.

13

CD-ROM Labels and Packaging

You can buy Avery CD-ROM labels and use the same process as discussed earlier. The selection is limited and not as professional as if you use software and labels designed for CD-ROM production. I use a dedicated software utility called MediaFACE that came with a CD label kit made by Neato. The CD-RW burning software that came with my computer also has a built-in label-making feature called Adaptec CD Creator. I will use the MediaFACE product to demonstrate the process here.

1. Open MediaFACE. The main screen will appear, as shown in Figure 13-2.

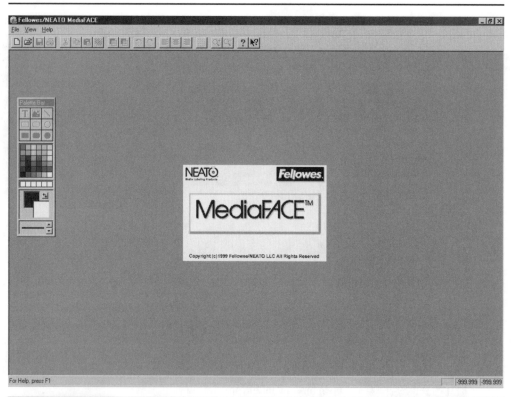

MediaFACE main screen

2. Select the label type in the dialog box shown below. I have selected the two-up CD disc label.

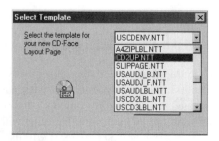

The label design template will open with a floating Palette Bar on the screen, as shown in Figure 13-3.

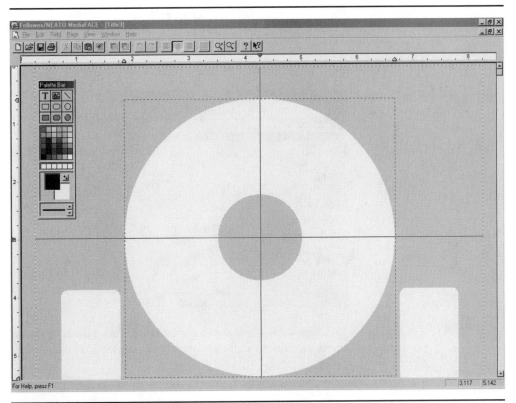

FIGURE 13-3   Template screen with Palette Bar

3.   Click the text button (the T on the Palette Bar). This will open the Text Field Properties
     dialog box shown in the following illustration. If it is not already chosen, click the
     General tab. Enter the text for a text field to be applied to the label. You can apply as
     many text fields as you need to make your label.

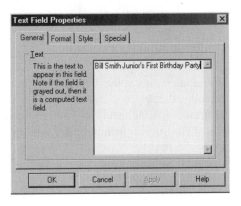

4. The next tab to choose in the Text Field Properties dialog box is Format. Click Change Font to bring up the Font dialog box shown next. Here you can make changes to text properties such as font, size, color, and so on. Click OK when you're satisfied with your selections.

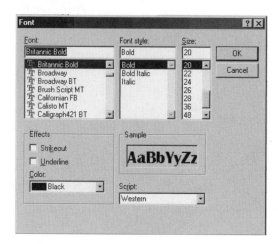

5. The next tab is Style, shown in the following illustration. This will allow you to choose a flat, curved, or other style of box to apply to your label. I have selected Curved and accepted the default.

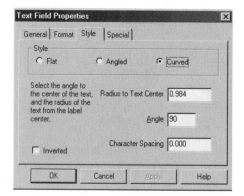

6. The last tab is Special, shown in the illustration on the following page. This will allow you to include automatically generated information about your project, like clip lists and copyright notices, on the labels and other printed material. I won't cover this feature here, as it's a specific feature of this software when used with a CD-RW device.

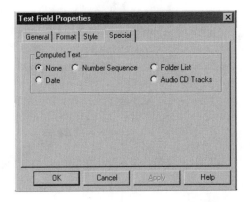

7. When you have made your choices in the tabs, click OK to add the text field to your label (see Figure 13-4). You can add more fields as required.

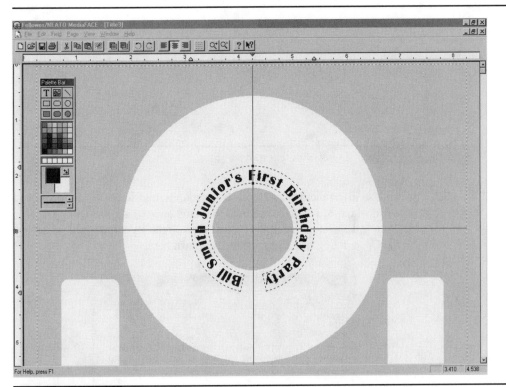

FIGURE 13-4   Label template with text added

8. In Figure 13-5, I have dragged the text field to the outer edge of the CD. You can change the text placement at will.

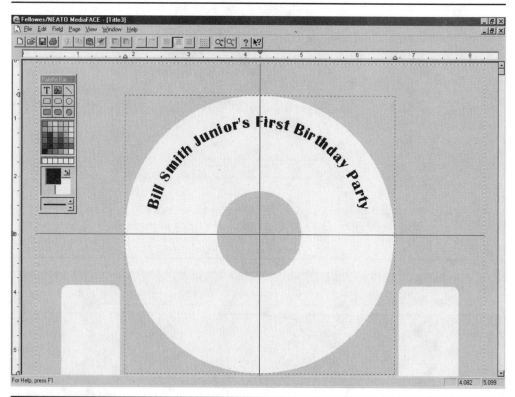

FIGURE 13-5   Label with text moved

9. You can also insert pictures, textures, backgrounds, and so on. In short, you can create virtually any kind of label you are used to seeing on your audio CDs or DVDs from the store. To insert a picture, click the Picture icon on the Palette Bar and select a picture from the Open dialog box that appears, as shown here:

10.   I added a graphic of the seacoast where Bob Junior's birthday party took place (see Figure 13-6). I have also changed the properties of the text to white from black so it will show up on the label. When you have finished designing your label, follow the printing and application instructions for your printer and the label.

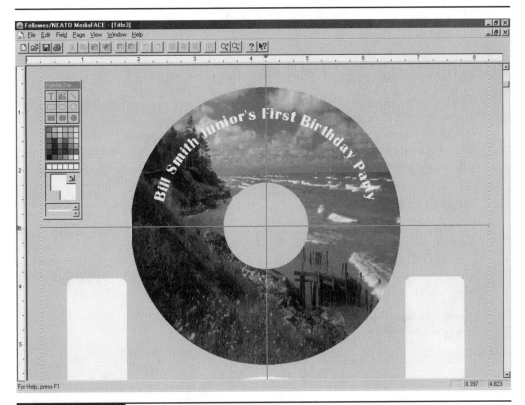

FIGURE 13-6   Label with picture added

13

You might want to create jewel box labels for your CD as well. Use the proper templates and the same procedures outlined earlier.

Copyrights and Copyright Notices

You will want to include a copyright notice on the package or label of your videotape or CD-ROM as well as within the credits and titles. Typically this is done on the last section of credits or on a separate title slide at the end of the production.

Copyright use is a complicated subject, and when I teach video production a great many questions on copyright always come up. I have covered the primary questions that arise in this section.

 *A basic copyright notice will consist of: Copyright © 2001 (use the actual year you create your project), Your Name, and All Rights Reserved. This will cover most situations.*

Understanding and Clearing Copyrights

The following information was derived from a much more complete guide available from the U.S. Library of Congress website at http://www.loc.gov/copyright. I make no claims to be offering legal advice in these matters. Please check with an attorney for specific requirements that may relate to your use of copyright material belonging to others and for copyright protection for your own works.

What Is Copyright?

Copyright is a form of protection provided by the laws of the United States (Title 17, U.S. Code) to the authors of "original works of authorship," including literary, dramatic, musical, artistic, and certain other intellectual works. This protection is available to both published and unpublished works. Section 106 of the 1976 Copyright Act generally gives the owner of copyright the exclusive right to do and to authorize others to do the following:

- To reproduce the work in copies or phonorecords

- To prepare derivative works based upon the work

- To distribute copies or phonorecords of the work to the public by sale or other transfer of ownership, or by rental, lease, or lending

- To perform the work publicly, in the case of literary, musical, dramatic, and choreographic works, pantomimes, and motion pictures and other audiovisual works

- To display the copyrighted work publicly, in the case of literary, musical, dramatic, and choreographic works, pantomimes, and pictorial, graphic, or sculptural works, including the individual images of a motion picture or other audiovisual work

- In the case of sound recordings, to perform the work publicly by means of a digital audio transmission

In addition, certain authors of works of visual art have the rights of attribution and integrity as described in section 106A of the 1976 Copyright Act. For further information, request Circular 40, "Copyright Registration for Works of the Visual Arts."

It is illegal for anyone to violate any of the rights provided by the copyright law to the owner of copyright. These rights, however, are not unlimited in scope. Sections 107 through 121 of the 1976 Copyright Act establish limitations on these rights. In some cases, these limitations are specified exemptions from copyright liability. One major limitation is the doctrine of "fair use," which is given a statutory basis in section 107 of the 1976 Copyright Act. In other instances, the limitation takes the form of a "compulsory license" under which certain limited uses of copyrighted works are permitted upon payment of specified royalties and compliance with statutory conditions. For further information about the limitations of any of these rights, consult the copyright law or write to the Copyright Office.

Who Can Claim Copyright?

Copyright protection subsists from the time the work is created in fixed form. The copyright in the work of authorship *immediately* becomes the property of the author who created the work. Only the author or those deriving their rights through the author can rightfully claim copyright.

In the case of works made for hire, the employer and not the employee is considered to be the author. Section 101 of the copyright law defines a "work made for hire" as:

- A work prepared by an employee within the scope of his or her employment

- A work specially ordered or commissioned for use as:

  - A contribution to a collective work

  - A part of a motion picture or other audiovisual work

  - A translation

  - A supplementary work

  - A compilation

  - An instructional text

  - A test

  - Answer material for a test

  - A sound recording

  - An atlas

- A work agreed to by all parties in a written instrument signed by them

The authors of a joint work are co-owners of the copyright in the work, unless there is an agreement to the contrary. Copyright in each separate contribution to a periodical or other collective work is distinct from copyright in the collective work as a whole and vests initially with the author of the contribution.

13

Two General Principles

■ Mere ownership of a book, manuscript, painting, or any other copy or phonorecord does not give the possessor the copyright. The law provides that transfer of ownership of any material object that embodies a protected work does not of itself convey any rights in the copyright.

■ Minors may claim copyright, but state laws may regulate the business dealings involving copyrights owned by minors. For information on relevant state laws, consult an attorney.

How to Secure a Copyright

The way in which copyright protection is secured is frequently misunderstood. No publication or registration or other action in the Copyright Office is required to secure copyright. There are, however, certain definite advantages to registration.

Copyright is secured *automatically* when the work is created, and a work is "created" when it is fixed in a copy or phonorecord for the first time. "Copies" are material objects from which a work can be read or visually perceived either directly or with the aid of a machine or device, such as books, manuscripts, sheet music, film, videotape, or microfilm. "Phonorecords" are material objects embodying fixations of sounds (excluding, by statutory definition, motion picture soundtracks), such as cassette tapes, CDs, or LPs. Thus, for example, a song (the "work") can be fixed in sheet music ("copies") or in phonograph discs ("phonorecords"), or both.

If a work is prepared over a period of time, the part of the work that is fixed on a particular date constitutes the created work as of that date.

Summary

I've covered a number of aspects of printing and packaging your project and appropriate copyright information for your packaging, tapes, and labels. This may seem to be unimportant to your video production, but it can make the difference between a "professional" result with accolades from your audience and embarrassment. Remember: don't drop the ball at the finish line. Pay attention to the final details of packaging and labeling and all of the other collateral material you might produce to promote your video project. Good luck at the awards banquet!

Part III

Take Your Edited Video on to Other Sources

Chapter 14

Transfer Your Finished Video to Tape and CD

How to...

- Transfer your edited digital video program back to your camcorder
- Export MPEG files from MainActor
- Burn a CD with digital video

Creating useable tapes and CDs of your edited video programs is a primary objective of most video editors. There are, of course, other uses for digital video, which I will cover in Chapter 15 and beyond. First, let's cover the basic use of your finished and edited video programs.

Transferring Your Finished Video Back to Your Camcorder

Most video editing programs provide a function on the menu or a separate program to control the process of transferring the DV files that have been created during the editing process back to the camcorder or other device for storage on DV cassette tapes. MainActor provides a program called DV-Out, a utility that allows you to transfer edited video created and saved in DV format with MainActor to DV devices such as a camcorder or digital VCR.

To use the DV-Out program, you will need the following items:

- A digital device such as a camcorder
- Windows 98 SE (ME) or Windows 2000
- A FireWire card and appropriate cable

 *The time it takes to transfer a digital video from a computer to a camcorder will be exactly the same time as it takes to play the tape. A 30-minute program takes 30 minutes to transfer.*

Overview of the DV-Out Window

The DV-Out main screen (shown in Figure 14-1) contains commands, settings, and a toolbar used for DV capture. The files you want to export must be in DV format. Before you can export them from the Sequencer you need to choose the MainActor export modules with the format AVI, and the codec MainConcept DV-Softcodec.

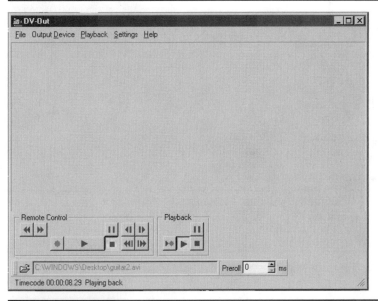

FIGURE 14-1   DV-Out main screen

The DV-Out Menu

In order to transfer your video from the computer to your camcorder or other device, you must set up several functions in the transfer software. The DV-Out menu provides access to the functions required.

1. Set the path and file name of the file that will be sent to the DV device by selecting File | Open, which opens the Open dialog box shown in Figure 14-2. Select the file you want to transfer. Make sure that it is in the proper AVI and MainConcept DV-Softcodec format. Click open.

2. Click Output Device in the menu bar. The Output Device menu that appears allows you to select the DV device (or a file) for output, as shown in Figure 14-3. You may have one or more DV devices connected to your computer. Select the one that you want to transfer the file to.

3. When you have made the proper selection you may start and stop the recording using the transport control buttons shown in Figure 14-4 or if you have a remote control function, your camcorder or other DV device will begin recording and transferring the file from your computer to the DV tape. You can view the resulting file from your camcorder viewfinder or an external monitor.

14

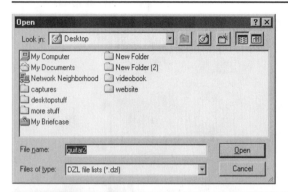

FIGURE 14-2   Setting the path and file name for transferring to the DV device

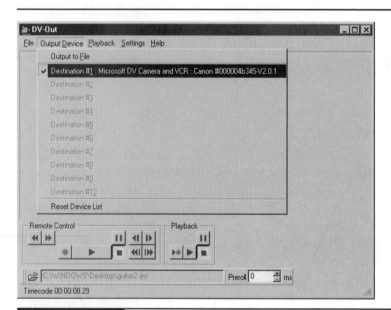

FIGURE 14-3   Selecting the DV device

If you have multiple FireWire ports, you may have more than one device attached to your computer, but you can only operate one DV device at a time over the FireWire connections.

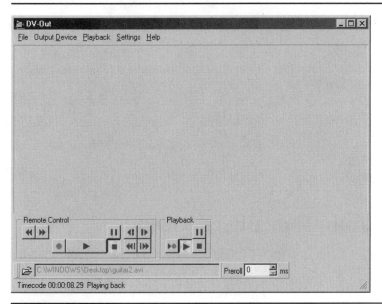

FIGURE 14-4   Begin the transfer process

The DV-Out Toolbar

The control buttons in the window can be used to control your DV device if your device supports remote control functions. Not all camcorders have remote control capabilities. If your camcorder or other device doesn't support remote control, you will have to manually set the tape position to the place where you want to begin recording. Check the manual for your device to determine if it supports remote control.

The transport control buttons are as follows:

- Rewind
- Fast-forward
- Pause
- Step forward by a single frame
- Step backward by a single frame
- Search forward
- Search backward
- Start playing the video
- Stop playing the video

14

The playback control buttons control the transfer of the video file from the computer to the DV device. The playback buttons are modeled to look like the ones you are used to using on a camcorder or VCR.

The right arrow play button with the red record button is for transferring the AVI file to your camcorder. The right arrow play button on the left side without the red record button starts playing the video without recording to tape.

There is no preview pane provided in the DV-Out window. You can use the camcorder display to review and monitor the transferred video.

You can monitor the progress of the video transfer in the viewfinder of your camcorder.

Creating a CD from Your Digital Video

To create digital video playable from a CD-ROM and to burn a CD-RW, you will need the following:

- MainActor, Adobe Premiere, or other video editing software
- A CD-RW drive in, or attached to, your computer
- CD-RW software such as Adaptec Easy CD Creator
- A CD-RW blank disc

Proper CD playback of MPEG-I files depends on the speed of the computer running the playback software and the speed of the CD-ROM drive. Some older computers may not support CD playback with high quality.

You will need to export your program or file from the DV format that it was initially in when you transferred it from your digital camcorder into MPEG-I format for normal playing from CD-ROM. There are other MPEG formats, such as MPEG-II and DVD, which have a higher resolution. These formats will not necessarily play from a normal CD-ROM and may require the use of special software and even special DVD recording hardware for the creation of a special video CD. You can find information about these and other formats by searching the Internet. This section covers only the standard MPEG-I and normal CD-RW formats.

Many computers today come equipped with a CD-RW drive, which will allow you to record any kind of data to a CD-RW disc. If your computer came equipped with such a drive it will already have a software package to take care of the recording process. If your computer didn't come with a CD-RW device, you can purchase either an internal or external drive that

will work with most computers. These drives cost between $100 and $300 and come equipped
with the appropriate software for recording.

*CD-ROMs created on Windows machines will generally work on Macintosh
computers. MPEG-I files should playback on either Windows or Macintosh
machines.*

Exporting MPEG-I Files from MainActor

You will need to open your MainActor Sequencer module and select the program or file you
want to place on CD-ROM. This is done by selecting File | File Open. The program will open,
as shown in Figure 14-5.

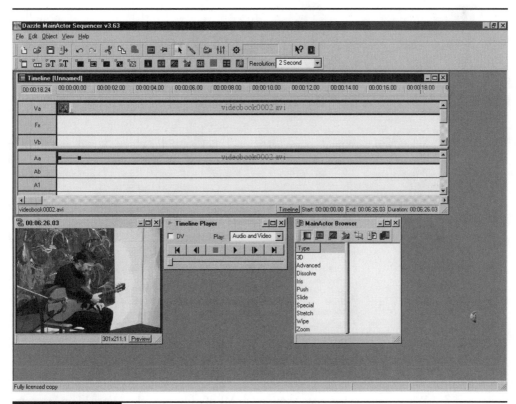

FIGURE 14-5   Select program to export to CD-ROM

14

Once you have loaded the file you will need to export the file from the DV format it was created in to the MPEG-I format for creation of your CD-ROM. When you select File | Export, the Export dialog box will open and allow you to select the export options.

1. Select the MPEGI/II export module and enter an Export Width of 352 pixels and an Export Height of 240 pixels. Export at 30.000 frames per second. The Export dialog box should look like this:

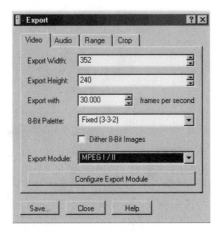

2. When you have made these selections, click Save and choose a path and file name to export your file to in the Select File dialog box, shown in the following illustration. Click Save.

3. When your file has been exported to MPEG-I format, close your MainActor module and you are ready to create your CD-ROM.

Embedding Your MPEG Movie in a Menu

You can use any HTML editor (HTML is Hypertext Markup Language, the format that web pages are created in) to create a simple web page menu to provide an interface for playing your MPEG movie from your CD-ROM. This web page can be copied onto your CD-ROM along with your video files and can control the playback of as many files as you want to link. A page can also be created for each file and a master menu page can be created that links your individual video play pages. You can include information about your video along with the video controls. I will cover this process in more depth in Chapter 15. In this case, let's use Microsoft Word to create a basic single menu page.

1. Open Microsoft Word and select File | New. Select Web Page as your page type. This can also be done by creating a standard Word page and then selecting File | Save As and saving the page as an HTML page.

2. Once you have your HTML page, add your basic information you would like to appear on the web page about your video clip or clips. This might include author, subject, date, copyright, and related information (see Figure 14-6).

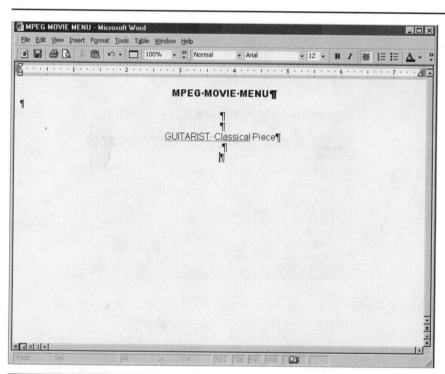

14

FIGURE 14-6   Microsoft Word HTML page with text added

3. Next, you will embed the control object that will play your video clip or clips. Select Insert | Object. The Object dialog box opens, as shown here:

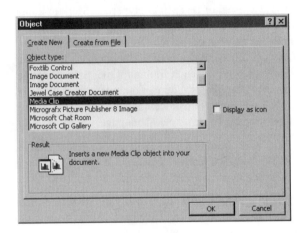

4. Select Media Clip and click OK. This will place a small square icon on your HTML page (see Figure 14-7). You may change its location as you desire. You will notice that the top menu bar changed when the media clip object was inserted.

5. Select Insert | Clip and then the MPEG option from the top menu bar. The Open dialog box appears, as shown in the following illustration. Select the MPEG movie clip you want to link to the control and click Open.

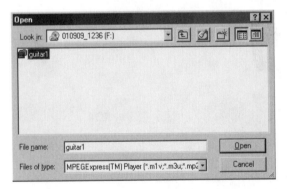

This will create a playback window for your video and insert a still image from the beginning to indicate the video, as shown in Figure 14-8.

6. Save the page under an easily identified name like videomenu.doc.

7. Copy the menu along with your video to the CD-RW using the procedures outlined in the following section. When you double-click on the videomenu.doc page from the CD, it will bring up the page for you to play your video or video clips.

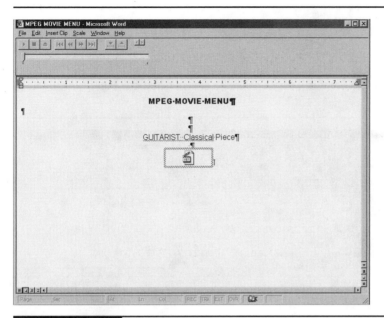

FIGURE 14-7   Embedded control object icon

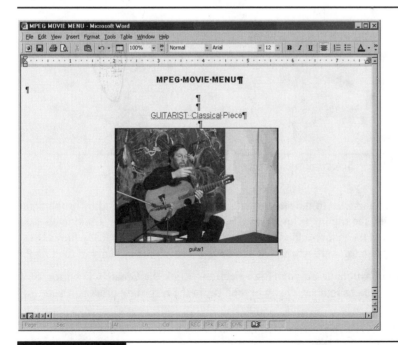

FIGURE 14-8   Inserting a still picture in the menu page

14

Burning Your CD

This demonstration will use Adaptec Easy CD Creator, but you can use your CD-RW recording software.

1. Open Easy CD Creator. The file selection screen (shown in Figure 14-9) will appear.

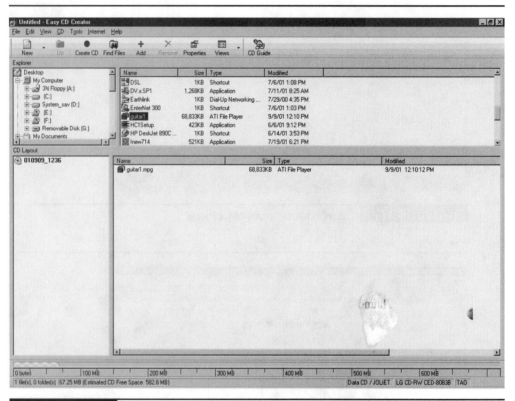

Selecting files to record to your CD-RW disc

2. Select and add files to the staging area on the lower-right window of the screen by highlighting the file in the upper-right window and clicking the + button on the toolbar. The selected file is then added to the list below. You can add as many files as you want until the program notifies you that you have exceeded the capacity of the CD-RW disc.

3. Once you have made all your file selections, click the Create CD button on the Easy CD Creator toolbar. This will then begin the recording process, as shown in Figure 14-10.

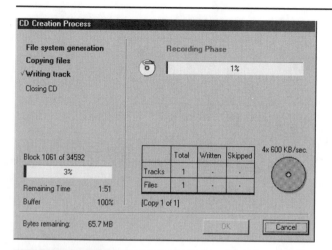

CD Creation Process

File system generation
Copying files
√ Writing track
Closing CD

Recording Phase

1%

Block 1061 of 34592
3%

	Total	Written	Skipped
Tracks	1	-	-
Files	1	-	-

4x 600 KB/sec.

Remaining Time   1:51
Buffer   100%

[Copy 1 of 1]

Bytes remaining: 65.7 MB

OK   Cancel

FIGURE 14-10   Recording progress box for your CD-ROM disc

Name your video clips with names that can be easily identified so you will know what to expect when the clip is selected.

When this process has occurred successfully, you can use the Windows File Manager to select the video file on the CD-RW disc you want to play. If your computer is configured to automatically play multimedia files—and most are—you can simply double-click the file icon and it will `nch the video player to view your MPEG video.

Summary

In this chapter you have learned how to transfer finished video back to your digital camcorder or other digital device. You've also learned how to create an MPEG video that can be transferred to a CD-ROM for distribution of your video programs as well as the process of burning a CD-ROM. There are many other uses for your digital video programs beyond DV tape and CD-ROM. You can copy from your camcorder to a standard VHS VCR tape and share your program with friends or you can create Internet video files and email them to associates. I will be covering Internet video in the next chapter.

14

Chapter 15

Use Your Video on the Internet

How to...

- Process your video for emailing, downloading, or streaming
- Create webcasts of special events
- Select and connect your equipment for webcasting
- Shoot the best video for use on the Internet

The Internet is one of the most compelling media through which you can share your video projects and use your digital video equipment. With no additional equipment and through the use of freeware software programs you can email video to friends and clients, create downloadable video (video clips that are downloaded to your hard drive and then viewed), or create streamable video (video that can be viewed over the Internet while it is being downloaded). You can even webcast an event, such as a church program, school meeting, lecture, or Bob Junior's first birthday party so that grandma back east can virtually be there through the Internet. This chapter covers processing your video for use on the Internet, whether for distribution, use in presentations, or to stream from your website. It also covers the basic concepts of webcasting.

> **NOTE** *DSL lines or cable modems make sending and receiving video fast and convenient. A word of caution, however: As video files are large, you should ask the recipients before you email a video, in case the file is too large to receive conveniently. If you are offering video on your website, either streamable or downloadable, the viewer will have a choice of whether they download it or not. You might consider adding the file size next to the download link on your web page.*

Creating a Video for Downloading

While it is possible to download any video file over the Internet, raw DV files are 4 MB per *second* of video, and it is impractical to offer these as downloadable files or email them to friends. The solution is to compress the files to a much smaller size using MPEG-I or other formats designed for Internet delivery. To create digital video over the Internet, you will need MainActor, Adobe Premiere, or other video editing software.

You will need to export your program or file from the DV format it was in when you transferred it from your digital camcorder into MPEG-I, RealVideo, or Windows Media format in order for the file size to be small enough to download practically over the Internet. There are other MPEG formats such as MPEG-II and DVD, which have a higher resolution, but these files also have a much larger size than MPEG-I. You can find information about these and other formats by searching the Internet. I will cover only the standard MPEG-I and RealVideo formats here.

Exporting MPEG-I Files from MainActor

First, open your MainActor Sequencer module and select the program or file you want to place on CD-ROM, as shown in Figure 15-1.

Once you have loaded the file, you will need to export the file from the DV format it was created in to the MPEG-I format for creation of your CD-ROM.

1. Select File | Export. This will open the Export dialog box shown here:

2. Select the MPEGI/II export module, and enter an Export Width of 352 pixels, an Export Height of 240 pixels, and Export With rate of 30.000 frames per second.

3. When you have made these selections, click Save. The Select File dialog box appears, as shown here:

4. Choose a path and file name to export your file to.

5. When your file has been exported to MPEG-I format, close your MainActor module and you are ready to create your CD-ROM.

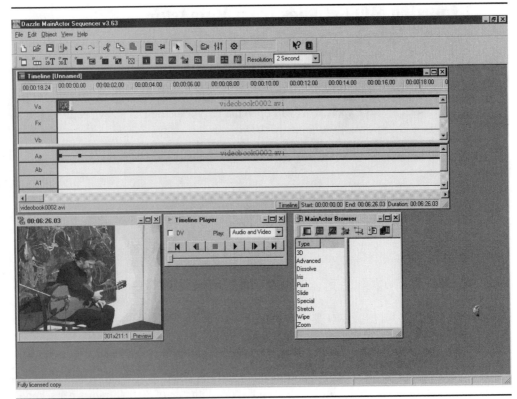

FIGURE 15-1   Select program to export to MPEG-I

Creating a Video for Streaming

To create RealVideo files or Microsoft Windows Media files for streaming, you can use the same process described previously, selecting the RealVideo or Windows Media export modules as your output file type.

Another way to create streaming media files is by using RealNetworks Producer Basic or Producer Plus, which will translate your DV file into the appropriate RealVideo file as well as help you create web pages with embedded video. Later in the chapter, I'll cover how you can also use the Producer program to stream your video over the Internet in real time.

Using RealNetworks' Producer to Create Streaming Video Files

RealNetworks Producer Basic is a freeware product available at http://www.realvideo.com/. Its more advanced sibling, Producer Plus, with many more features, is available for purchase at the same website. For creating simple and basic streaming video and embedded web pages, the Basic product will do just fine.

1. Download or purchase RealNetworks Producer Basic or Plus.

2. Install and start Producer. The opening screen appears, as shown in Figure 15-2. Soon after, a wizard starts automatically and walks you through the process of importing and converting a DV file.

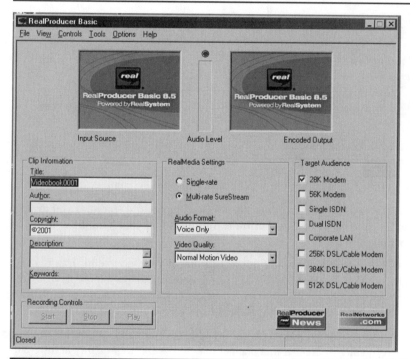

FIGURE 15-2   Opening screen of Producer

15

3. In the first wizard screen that opens, shown in Figure 15-3, select the Record From File button. This will allow you to select and convert a prerecorded video clip in a number of different formats into a RealVideo file. The other two choices allow you to create a RealVideo file from a live camera for distribution later and to webcast live in real time. We will not cover these two options.

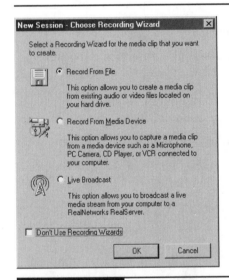

FIGURE 15-3 Choose Recording Wizard screen

4. The first screen in the Recording Wizard (see Figure 15-4) prompts you to select a video file you want to process. Browse for this file on your hard drive, then click Next to continue.

5. You will then be asked to enter information about your video clip. This information will be stored in the RealVideo clip itself and will be displayed on playback to inform the viewer what the clip is, who the copyright holder is, and so on (see Figure 15-5). Click Next.

6. The next screen allows you to choose whether the file will be restricted to a specific playback quality or created in a manner that allows for high quality at different modem speeds (see Figure 15-6). The Multi-Rate SureStream for RealServer G2 choice is recommended. Click Next.

7. You will then need to choose the types of modems your viewers are most likely to have. If you have doubts, choose the 56K and 256K DSL/Cable Modem choices (see Figure 15-7). This will cover most home users. If you are using Producer Basic, you will only be allowed two choices. Click Next to continue.

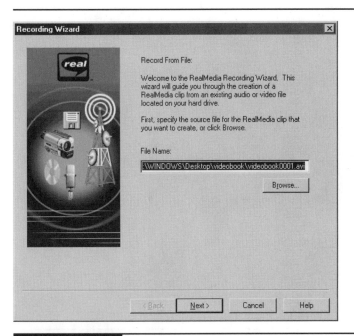

FIGURE 15-4   Select a video file

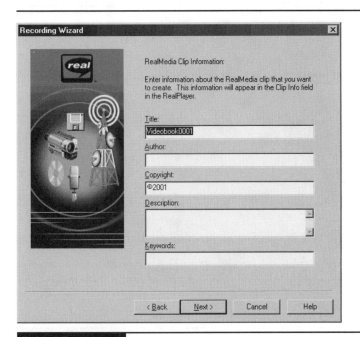

FIGURE 15-5   RealMedia clip information screen

15

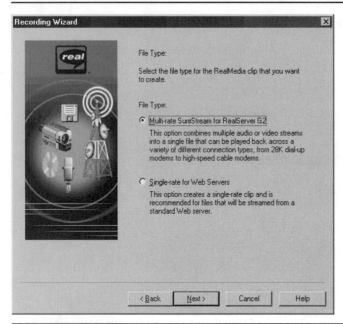

FIGURE 15-6   Select output file type

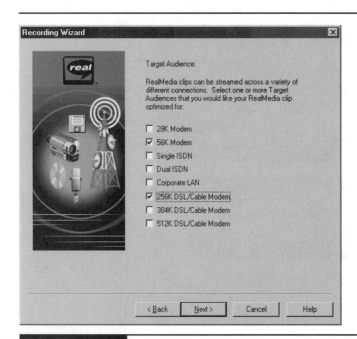

FIGURE 15-7   Select target audience

8. Your next choice will be the audio format (see Figure 15-8). It ranges from voice only to stereo music. Choose the one that most closely describes your video clip's audio track, and click Next.

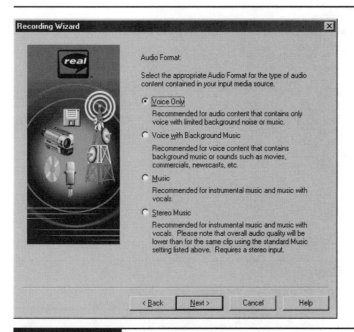

Select audio format

9. The next screen prompts you to choose the video quality (see Figure 15-9). Choose the lowest quality that you feel you can get by with. This will accommodate the slowest modem you will encounter. Click Next.

10. You will now need to choose a file name and path to store your RealVideo file (see Figure 15-10). Click Next to continue.

11. When you have made your selection of video quality, the next screen presents you with all your choices, as shown in Figure 15-11. You can back up and make changes or you can click Finish to begin the conversion process.

12. A new screen will appear that presents the initial frame of your video and a second screen to display the converted video file (see Figure 15-12). It will also allow you to make final adjustments in the parameters. When you are ready to create your file, click Start. The screen will look like Figure 15-13 during the processing.

15

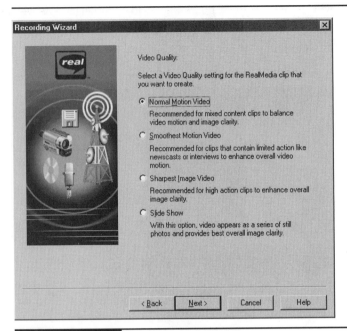

FIGURE 15-9   Select video quality

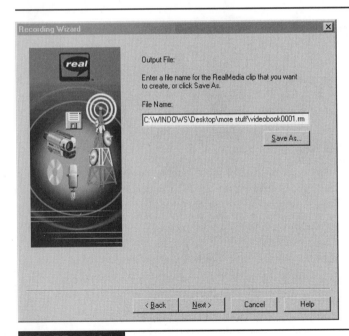

FIGURE 15-10   Select output file name

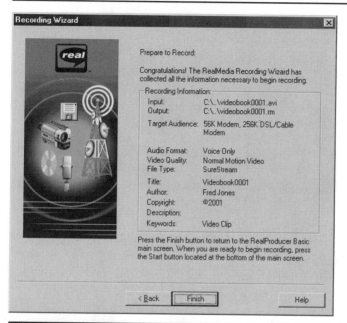

FIGURE 15-11   Prepare to record

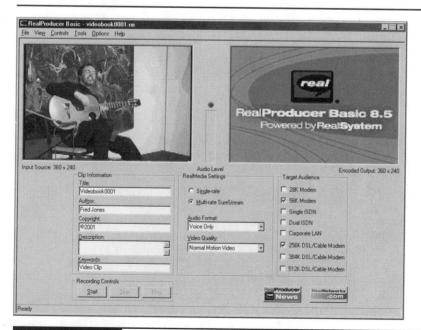

FIGURE 15-12   Select Start button to create file

15

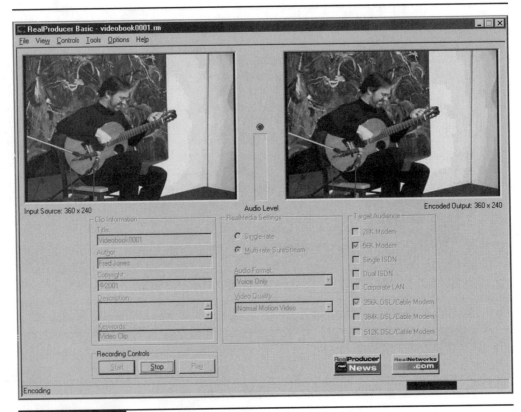

13. When the process is finished, a window will pop up telling you the status, as shown here:

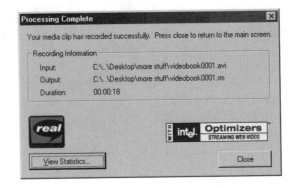

The final screen will look like Figure 15-14. This indicates the final properties of the file and notifies you that the processing is complete.

FIGURE 15-14   Preview your file

14. When the file conversion is finished, you may preview your finished file. If it isn't satisfactory, change the settings and process it again until you get it right.

15. When the file is finished, you can exit the program and send the file as an email attachment or manually add it to a web page.

If you want to use the web page creation feature of Producer, select the Tools | Create Web Page option from the menu bar. This will bring up the Web Page Wizard, which will walk you through the process of creating a web page with your video clip embedded in it (see Figure 15-15).

15

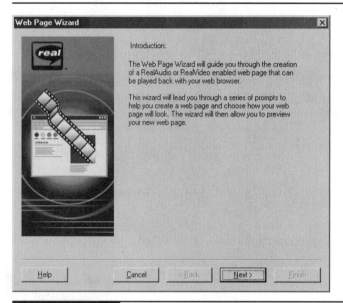

FIGURE 15-15   Web Page Wizard

1. Select the video file you want to create a web page with (see Figure 15-16). It will default to the one you just created or you can choose another or come back later and create one or more web pages.

2. The next screen offers you a choice of two playback methods (see Figure 15-17). Choose the Pop-Up Player. Click Next to continue.

3. On the next screen, shown in Figure 15-18, enter a caption that will describe your video clip on your web page. Click Next.

4. Choose the web page file name and save path as shown in Figure 15-19, then click Next in order to proceed.

5. On the final wizard screen, shown in Figure 15-20, you can preview and make any changes to your settings. Click Finish when you have reviewed your choices.

6. Your web page will be created. You can now go the directory it is stored in and view it with your web browser.

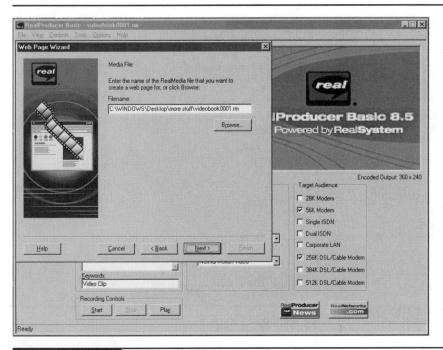

FIGURE 15-16 Select video file

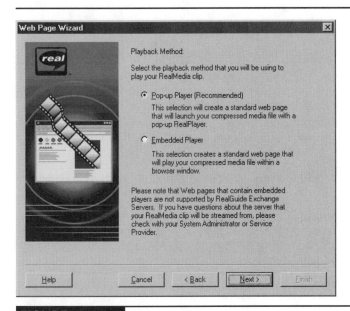

FIGURE 15-17 Choose playback method

15

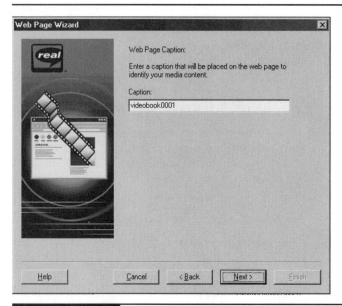

FIGURE 15-18   Enter a video caption

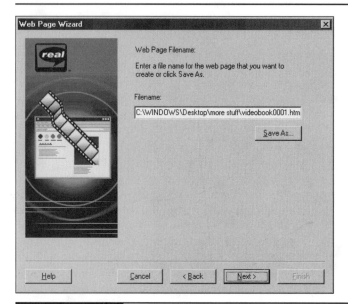

FIGURE 15-19   Choose the web page file name

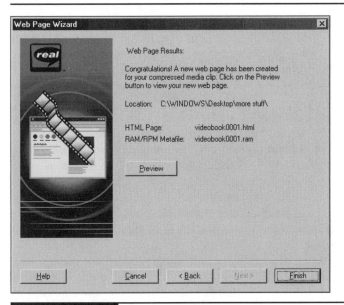

FIGURE 15-20   Preview your finished web page

Webcasting

Webcasting is not about technology, rather, it's about content and communication. Communication only occurs when someone responds, and this in turn is the secret of content. When you communicate, you create your own content. For example, you can tape an hour-long telephone conversation (as in a telephone interview with an expert or notable personality), then edit and assemble it into a good half-hour radio program by inserting an introduction, and bridging and closing narration together with musical bridges and sound effects. The same applies when you are using video conferencing. The trick is to recognize the potential of the content, exercise good judgment, and build your production values into the communicating process. You can dance, play music, create dramatic productions or public affairs programs suitable for broadcast and cablecast in shared electronic space, providing you record or tape simultaneously at each location and have someone to edit and assemble the final production.

You can also create content with webcasting and the Web specifically in mind, such as original games or interactive fiction. It can be drawn from other formats or sources, such as radio or television broadcast, or cablecasts that are repackaged for the Internet. You can create content from an original design or capture it as an event unfolds, as in news event webcasting or collaborative educational webconferences. The rules for webcasting content are being made up as we all go along. The early stages of most new media are flooded with recycled content from other, established media. The first movies were often films of stage plays. Seminal television

15

consisted of broadcasted radio or vaudeville programs. Quickly, however, the inherent strengths and weaknesses of any new medium dictate both the form and content delivered through them. Ultimately, their audiences establish a demand for certain popular and successful categories. The Internet began with text-only pages, located with the greatest of difficulty, by expert computer scientists. Now the Internet is an increasingly democratic medium, affording entry and production to small business and individuals, where once only mega-corporations had the facilities to produce content for mass consumption.

There is a flood of original educational distance-learning programming that can be categorized as webcasting. It includes a diverse range of projects from the delivery of real-time online classes to the distribution of classroom support material similar to cable or broadcast educational television. These programs exist for K–12, technical education, self-improvement, continuing education, corporate training, and undergraduate and graduate degree programs. Many of the online educational sites are little more than email clearinghouses, but there is a major trend toward the inclusion of audio and video teaching resources, as well as real-time classroom and webconferencing. Since the Internet was invented to support education, this trend is both natural and exciting in its scope.

Webcasting is also practical for events that would be inappropriate for traditional radio or television broadcasting because of audience size, cost of distribution, or the content of the event. Examples range from the webcasting of musical or comedy concerts of new and emerging artists to extensive detailed coverage of protracted and complex events like elections or a NASA Mars landing.

Event coverage is sometimes a function of news coverage, but can be broadened to include entertainment and educational events as well. Trade shows, conferences, conventions, rallies, sports, and many other events are candidates for the enhancement of webcasting. Major events that garner network and cable television coverage can be expanded with background information. In other cases, where the event is not large enough or appropriate for that kind of attention, primary webcast coverage is feasible technically and financially. In a different venue altogether, the webcast coverage of new musical groups, for example, can become a major marketing tool. Corporations can establish press conferences and training sessions for new products that are being introduced, and both employees and customers can be reached in much larger numbers and at much lower cost than ever before. This kind of corporate webcasting can validate the entire concept of webcasting and support its growth. The NASA Mars landing is an example of a major story, and its webcast coverage expanded the experience for millions of visitors to the website.

Tools for Webcasting

The tool that I most often use for webcasting is the RealSystem Producer Basic and its more complex siblings such as RealProducer Plus. RealProducer Plus converts live feeds or existing audio and video files into RealAudio and RealVideo files, ready for delivery on the Internet or corporate intranet. You have learned how to convert a DV file to RealVideo for downloading on the Internet as well as for streaming. Webcasting is simply the process of making the video available over the Internet. It can be done by shooting and editing the files for streaming later or can be done in real time with your camera hooked directly to the computer and the computer connected to the Internet.

Should You Webcast in Real time or Near Real Time?

You can tape your event, convert the files, and upload them to your website without much fuss or risk. I webcast an annual regional conference in this manner. We tape everything and convert the files to RealVideo in real time, then FTP the files to the website within an hour of the individual meetings. This serves our needs just as well as real-time streaming, and doesn't require an active RealVideo server and a fast Internet connection on site. This saves cost as well and the results are equally good for our purposes. An hour delay isn't even noticed by the outside world. I recommend this process, particularly for first time webcasters.

Occasionally you absolutely need to webcast in real time. Then the sense of immediacy can be effective, especially if you allow real-time feedback from web viewers. To webcast in real time, use the following process.

Setting Up a Project

If you are webcasting from a conference or meeting where you don't have as much control as you otherwise might, you should plan well ahead and test everything before the event begins. Here is a basic list of items you will probably need if you are webcasting in real time:

- A portable computer to run the RealVideo product software.

- An analog input on your computer or an add-on device such as the Dazzle Digital Video Creator to process your analog video into the computer.

- Your video camcorder.

- Cables to connect your camcorder to the computer. In this case you will need to connect the analog video and audio to an analog video and audio connector on your computer. You can get a feed to your camcorder from the event sound system if you think the quality will be better.

- An Internet connection to transfer your video to your RealVideo Server on the Internet.

- RealVideo Server installed on a server at the site or in another location.

You should shoot as you normally do, with Internet streaming in mind. I always record the video I am webcasting to tape, in case something goes wrong with the Internet connection or server and I need to process the video offline and upload it to the website later. This provision for backup provides a great comfort level.

Using RealVideo Server Software

You will need to link your portable computer at your event site with a server that is running RealNetwork's RealVideo Server software in order to webcast properly. This software is

15

provided on many website hosting services offered by ISPs and other providers. It is also commonly installed on business and institutional servers for internal use. Check with your ISP or the webmaster of your website for more information on the process and availability. If you maintain your own server, you can download a freeware version of the RealVideo Server and easily install it yourself, or you can purchase more expensive upgrades if you are expecting a lot of visitors to your site during the event. You should read the RealVideo Server manual carefully for bandwidth and other necessary information before beginning the process.

Always test everything well ahead of any event. Naturally, anything you don't try won't work! It's a rule of nature.

Transferring Your Files to the Website

If you are webcasting in real time, this will be done as part of that process. If you are uploading your files for later access, you will need to use FTP from your computer to transfer the files to your website. Check with your Internet service provider for details of the FTP process if you don't already know how to do it, as it varies from ISP to ISP.

How to Shoot for Webcasting

Digital video, at the lower resolutions required for Internet delivery, is sensitive to movements of the camera and subject. When one frame after another looks pretty much like the frame preceding it, Internet video can look great. But when things start changing rapidly, sometimes the computer is not able to keep up and the quality is sacrificed. This does not mean your video is limited to lock-down shots of talking heads, but it does mean that you must be conscious of the effects of camera and subject motion. Use them sparingly, but use them effectively.

Beware of fast pans and zooms. These stress many video compression schemes and can deteriorate otherwise good-looking video. If you must use them, use them boldly. It is generally not the big move that causes trouble, but rather the host of little background movements that deteriorate quality the most.

Summary

This chapter only scratches the surface of the use of digital video on the Internet. The improvement in bandwidth, allowing more and larger videos to be transmitted and viewed over the Internet, will increase these options multifold. There are many video sites cropping up on the Internet (check out www.bmw.com or www.iMovies.com for interesting Internet movies), and it is clear that the Internet will be a new medium for both professional and amateur video production. It is an exciting time to be experimenting with digital video and all its possibilities.

Part IV

Create Practical Digital Video Projects

Chapter 16

Create a Video
Inventory of
Your Home

How to...

■ Plan a video inventory of your home

■ Create a shot list

■ Shoot clips for your inventory video

■ Organize your clips

■ Create simple titles

■ Edit your inventory video

■ Create a final CD-ROM of your inventory video

Several years ago I received a phone call in the night from a friend. He informed me that my business offices were on fire and the firefighters were working on putting it out. I rushed down to the office and experienced one of the most emotional events in my life as I watched my business burn. I don't wish it on anyone. The next day the insurance adjuster showed up and we began to reconstruct the contents of the building in order to file a claim. It turned out that I didn't have good proof of much of the office contents, and much of what I had was in file cabinets in the charred office. The long and short of it was that I received a settlement of less than 80 percent of what it should have been. It was a severe loss to my business and my peace of mind. The long-term damages from this fire would have been much less severe if I had had good documentation and had stored it in a safe place at my home or in my bank safety deposit box. One way you can justify the price of your new digital camcorder is to create a video record of your home or office contents in case of a similar loss.

Although you can certainly make a list on paper and use a digital still camera or standard photographic camera images to create a visual record, doing it on video allows you to create an audio track and to show things in context, which can't be done with a still image. Recently a family member had a disastrous house fire. In order to process her claim, the insurance company insisted that she imagine herself walking, room by room, through her house, and listing every item and its location. It took her weeks to complete and was emotionally devastating. How much less harrowing it would have been if she had had a videotape of exactly that process safely in the bank. The same need may arise out of burglary and theft. Everyone needs an accurate and up-to-date inventory of their personal property. You can just shoot a videotape and take it to the bank without editing, of course. But the advantage of editing, beyond the opportunity to learn shooting and editing basics with a resulting valuable product, will be the process of thinking through what to include and how to document it. The process of planning and editing the video ensures completeness and will impress an adjuster should you ever need to provide documentation of a loss.

NOTE   *This chapter repeats things you might have learned in earlier chapters. It is redundant in order to enable you to fast-track a particular type of project using the following "cookbook" instructions. It doesn't, however, repeat the basics of using the camera and editing software. This information is contained in the first two sections of the book.*

Planning a Video Inventory of Your Home

The way I have chosen to organize such a video is on a room-by-room basis. You could also organize it by type of item (jewelry, furniture, art, and so on). The advantage of using the room-by-room approach is that your equipment will be in one place and it will ease the shooting process. It is also easy to visualize and remember the physical location of items if you ever have to use the video for an insurance claim. Most insurance adjusters assess fire and other damage to property on a room-by-room basis as well and your video could assist this process.

How to Record Various Items

The following will give you some hints for recording information about the various types of household items as you create your household inventory video. It is by no means exhaustive, but it should give you a start.

> **TIP**   *Remember, it's too late to do this after a disaster has occurred, so err on the side of too much information if there is any doubt.*

Recording Furniture

When you shoot furniture, you should get a good view of the furniture type and style as well as the general condition. If possible, you should indicate on the audio track the brand, year and place of purchase, and the approximate purchase price. Special notations should be made for antique furniture.

Recording Your Art Collection

If you have incidental prints, vases, decorative objects, and such, you can cover these lightly with a general shot. If you have particularly valuable paintings, prints, or antiques, you should shoot close-ups and details of each item individually. Antique furniture should be shot from the front, back, and even underneath if possible. Ceramics and similar items should have close-up shots of the bottoms with any markings that might indicate type and age. If possible, you should indicate on the audio track the brand or artist, year and place of purchase, and the approximate purchase price.

Recording Your Book Collection

Books can be shot as a general view of book shelves indicating the type and quantity. If you have antique or collectable books, they should be shot individually with front, back, and inside shots indicating condition. A shot of the title and copyright/edition page will also be important. If your editions are autographed, these should be shot and important information indicated. If possible, indicate on the audio track the year and place of purchase and the purchase price. If you have a large or significant collection, you should create a database or word processing file

16

inventory indicating all the key information. This can be stored on the CD-ROM created for the video.

Recording Your Electronic Equipment

Electronics can be shot as a group unless there are particularly valuable items. Indicate on the audio track the brand, serial number, model number, place of purchase, and the approximate purchase price.

Making an Inventory of Your Important Silverware

Silverware, particularly sterling silver, can be shot laid out on a dining table or inside a silver chest. A close-up of the back of a spoon or fork showing the maker and pattern will help. The audio track should indicate the number of place settings and significant items.

Making an Inventory of Your China and Crystal

China and crystal can be shot laid out on a table or within a china cabinet. A close-up of the back of a plate showing the brand and pattern will help, and the audio track should indicate the number of items and place settings. If possible you should also indicate on the audio track the year and place of purchase and the approximate purchase price. If you have a particularly valuable collection you should also create a database or word processing inventory to store on the CD-ROM you will create for the video.

Record Your Appliances

A general shot of the item is satisfactory, with make and model number. If possible, you should indicate on the audio track the year and place of purchase and the approximate purchase price.

Record Your Kitchen Tools and Equipment

A general shot of the inside of cabinets and drawers will give an adjuster an idea of the value of their contents. Include your kitchen utensils and cookware. If you have particularly expensive tools or items, you might want to take a shot showing their specific value and any other information and indicate it on the audio track.

Inventory Your CD and Tape Collections

Treat CDs, tapes, and records the same as books. If you have a valuable collection, you should shoot them in general and select key expensive items for close-ups. Maintain a print list or a database or word processing file with important information about each one including the title, artist, issuing company and date, purchase date and price, and so on. Describe anything special on the audio track.

Record Your Computers

Treat computers the same as electronics. You should create CD-ROM or tape backups of your important data and store them off site with your video inventory CD-ROM. Software is often as expensive or more expensive than the hardware. Remember to tape the boxes and/or master disks and record the product number, versions, and serial numbers on the audio track. It is a good idea to keep the master disks or CD-ROM copies of them in an off-site location.

Inventory Your Jewelry

Costume jewelry can be covered with a general shot of the items in their boxes or arrayed on a flat surface. Valuable items should be shot in close-up one by one with the audio track indicating type, date and place of purchase, purchase price, appraisal information, and anything else that will indicate value.

Inventory Your Clothing

A general shot of closets and drawers will suffice unless you have furs or other valuable items. These should be shot individually, including the linings and labels. Indicate on the audio track the brand, year and place of purchase, and the approximate purchase price.

Record Your Tools

A general shot of tools will be satisfactory. In the case of expensive specialty tools, be sure to include shots of the brand and model number where applicable. If you have an antique tool collection, treat it as you did other antiques. Note the date and purchase price of tools and equipment on your accompanying audio track.

Record Your Musical Instruments

If you own musical instruments, be sure the video includes the make. Include pictures of cases, and if model numbers are visible, include those as well. Otherwise, include this information, as well as the date and price of purchase, and age of the instrument, on your accompanying audio track.

Organization of Your Script

The basic script, in this case, is based on the rooms and type of contents that are typically contained in them. When you shoot, you will create a more detailed shooting list, but you will organize the shoot and subsequently edit the clips according to this scheme.

1. Living room

   Furniture

   Art

16

Books

Electronics

2. Dining room

Furniture

Silverware

China and crystal

Art

3. Kitchen

Furnishings

Appliances

Tools and equipment

Cookware and utensils

4. Family room

Furniture

Electronics

Art

Books/CDs/tapes

Games

Musical instruments

5. Utility room

Appliances

Tools

6. Office

Furniture

Computers

Art

Software

Office supplies

7. Garage

Tools

Appliances

8. Bedroom

   Furniture

   Electronics

   Art

   Jewelry

   Clothing

Creating a Shot list

You will now select one room from the script above and make a more detailed shooting list of the actual contents and the order in which you expect to record them. You will also create a sketch of the room, as shown in Figure 16-1, with the key items and the anticipated individual shots indicated and numbered. It will be a good thing to update any changes to the shot list when you are finished with each room. This list will then become an edit list that can be stored with the finished CD-ROM or tape to help document your inventory more fully for an insurance adjuster.

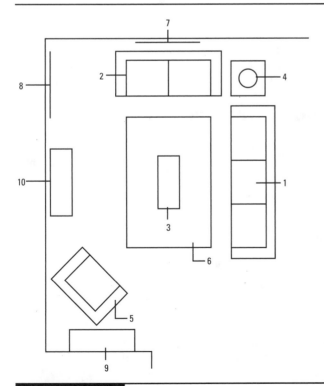

16

FIGURE 16-1   Sketch of living room and contents

Living Room (10 Clips)

Furniture

1. Sofa

2. Loveseat

3. Coffee table

4. End table and lamp

5. Easy chair

6. Oriental rug

Art

7. Painting on north wall

8. Paintings on west wall

Books

9. Bookcases

Electronics

10. Entertainment center

The same setup will be used for each room. You can number the clips sequentially for the entire video or begin the numbering again for each room.

Shooting Clips for Your Inventory Video

A video inventory isn't an art project. You just need to shoot accurate and clear video of the items of importance in your home. It is also important to record clear audio of your comments during the process. Try to keep the background noise and off-camera chatter to a minimum when you are shooting. This will make your finished video more useful in the future.

Set Up Your Equipment

The equipment you will need for this shoot may include:

Video camcorder and tape   This is the minimum equipment, of course. You can do an excellent job of recording your household possessions with only this basic equipment.

Tripod   In many cases, it may be easier to shoot much of the room and many of the items using a camcorder mounted on a tripod. You will have to think through the items and how they are best captured as well as what equipment you have available.

Extra lighting   If you are taping small items or your room has inadequate lighting, you might need a handheld or camera-mounted sun gun, a tripod-mounted photoflood, or just an extra lamp in the room. This will be more evident when you shoot closets, garages, and the inside contents of cabinets. Adequate lighting is absolutely essential—in the case of a claim, the clearer your evidence the better.

Copy stand   A copy stand is a device used most often by photographers to mount the camera above and perpendicular to objects to photograph them (see Figure 16-2). It can be very useful when taping books, jewelry, or art objects because the camera is fixed and the objects can be moved in place below it. You can also improvise one using your tripod.

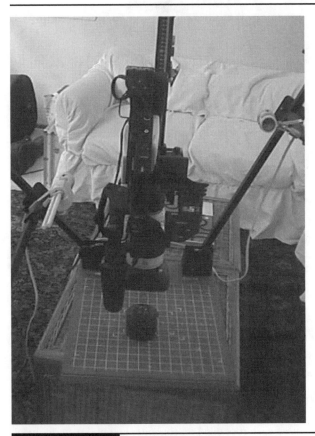

FIGURE 16-2   Copy stand with camcorder

16

Microphone mounted on a headset   A headset microphone attached to the external microphone jack of your camcorder will be very handy, especially if you are holding the camcorder in your hand rather than using a tripod, as shown in Figure 16-3. These microphones are easily purchased at Radio Shack or other electronic stores and often are the same items used as external devices for cellular telephones. Check with the salesperson for proper connectors for your camcorder.

FIGURE 16-3   Headset microphone

 *Take your camcorder with you to the store so the microphone jack and the microphone you purchase can be checked for proper connectors.*

Organizing Your Clips

Once you have shot the video clips of each room and its contents, you will want to update your shot list before you begin capturing and editing the clips. It will be easier if you updated your shot list as you captured your video clips, but it isn't too late if you didn't. Just sit and view the tape or tapes you recorded and make notes of each clip.

Note the start and stop times on your camcorder viewfinder, or turn on the time code display to show on your TV monitor plugged into the camcorder if your camcorder allows this display. Record the times on your shot list so it will be easier to locate each clip when capturing them to your hard drive. This is particularly important if your shots are out of order or if you have many false starts when recording.

Capture Your Video Clips to Your Computer with DV-Capture

Use the same procedure outlined in Chapter 8 to transfer your video from your camcorder to your computer. If you are using MainActor it will be identical. If you are using Premiere or another video editing software, adjust the procedure accordingly.

Creating Simple Titles to Separate the Room-by-Room Scenes

A title is the video element or clip with text that, in this case, will display the name of each room segment in your video. You can, if you wish, create separate titles for each item in your inventory or each group of items. In any case, use the following procedure. You are going to create a set of simple titles using the basic Windows Paint software utility. This is the easiest and most basic way to create titles.

1.   Under Start | Programs | Accessories in Windows you will find a utility program icon called Paint. Open the program. The main screen will appear as shown in Figure 16-4.

2.   Select Image | Attributes from the menu bar. The Attributes dialog box will appear, as shown in the following illustration. Type in the image size 352 pixels by 240 pixels. This will create a white background for your image the size of an MPEG video frame.

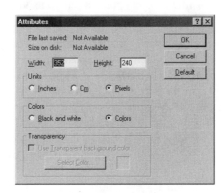

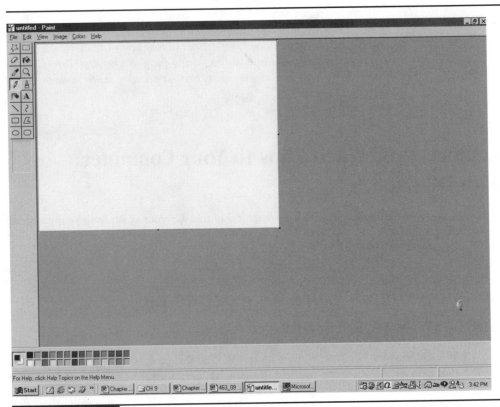

FIGURE 16-4   Paint main screen

3. Click the toolbar icon A. This tool enables you to insert text into your picture, which at this point is still a white blank background, as shown in Figure 16-5. A small crosshair will appear. You must select an area to contain your text. This is done by clicking the screen at the upper left of the image. Then drag the mouse to the right and down until you create a dashed square about 1/4 of the size of the screen. Don't worry if it is too small or large or in the wrong place when you add the text. This can be adjusted later.

4. Once your text square is created, the cursor turns into a large I beam when moved into the text square. This indicates that text can be placed if you click the mouse. Click and a blinking cursor will appear in the upper left of the text square. You may now enter your text. Type **LIVING ROOM** in all caps. Right-click within the text box and select Text Toolbar from the bottom of the menu box. This will open the Fonts toolbar, as shown in Figure 16-6. You can select the font, style, and size from this toolbar. I have set the attributes for Arial Black and the font size at 24. If you prefer a different text or background color for your title, you can adjust the attributes of your image accordingly. Refer to Paint's Help topics for information about using the utility.

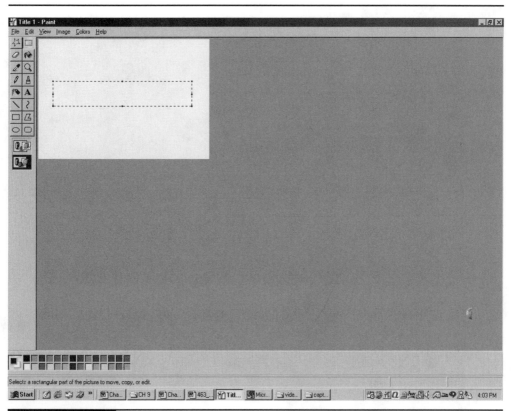

FIGURE 16-5   Inserting a text square in your image

5. Save your file in a format of your choice in your working directory for your project by selecting File | Save As from the menu bar and entering the file name in the Save As dialog box that appears, as shown in the illustration that follows. I selected the 24-bit bmp format, but any of the selections will work.

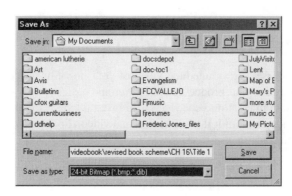

16

FIGURE 16-6   Fonts toolbar

6.   Create as many titles as you need for your project using the same procedure. Again, you
     may use titles for the rooms only or create titles for each category of items you are
     taping. The level of detail is up to you, but the basic use of titles indicating at least the
     rooms is really very helpful in clarifying your inventory video.

Editing Your Inventory Video

You may, of course, use an unedited video of your home or office contents, but editing your
inventory will make your product much clearer and easier to use. Besides, it's good practice!

     The first step in creating your video is to insert a video clip into the timeline. You will then
insert a title slide and turn it into a three-second video clip. The video clip will become the first
element of your video.

1. Start the MainActor Sequencer. You will see the main window pop up with the Video Profile window on top, as shown in Figure 16-7.

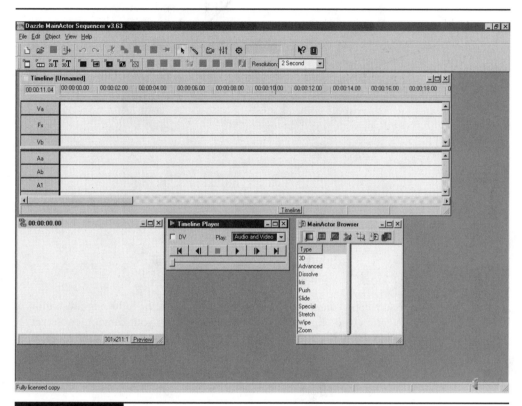

MainActor Video Profile window

2. Accept the Digital Video (DV-NTSC. 30 fps) from the Video Profile menu by clicking OK. You will then see the MainActor Sequencer main window. You can stretch and drag the various windows of the Sequencer into any order you want.

3. Place the cursor in the Va timeline and click. This will open a drop-down menu. Select Insert Multimedia as shown in the illustration that follows.

16

4. The Import Dialog will appear, as shown in the illustration that follows. This dialog box lets you select various import modules (small utility programs that convert and process video formats for integrated editing) that can be used to bring files into your project. In most cases the default settings are best. Click OK.

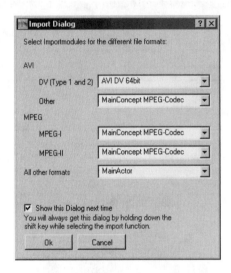

5. The Select File(s) dialog box will appear as shown in the following illustration, and you can select a video clip that you have captured from your camcorder. Select the file you want to import and click Open. (Refer to Chapter 7 if you need more details.)

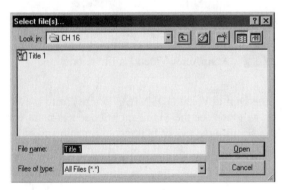

6. Move the cursor to the Va track. The cursor will change to a small film icon attached in the lower right indicating that you are ready to insert your title image. There will also be a gray rectangle in the track that indicates the length of the video clip, as shown in Figure 16-8.

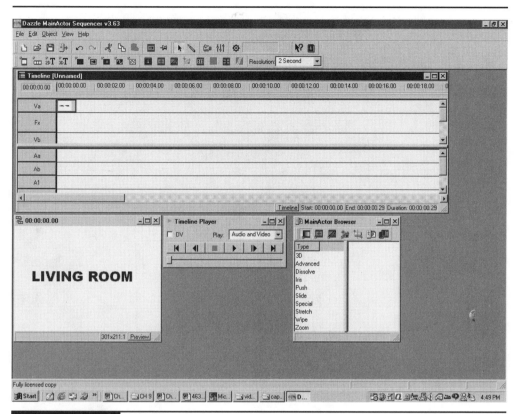

FIGURE 16-8   Inserting a title image into the Va track

7. Drag the clip with the cursor along the Va track until the timer reading is 00:00:03:00 as shown in Figure 16-9. This indicates that you are at the first frame of the title clip that you are creating. Click to finally place the video clip on the timeline.

8. Use the same procedure to insert video clips in the timeline. Insert title slides between each sequence of room clips. Continue adding video clips and titles to your timeline until they are all in place.

Unless it is a very short clip, you will want to change the resolution of the timeline from the default two seconds to a larger number. Use the Resolution pull-down menu on the toolbar to do this.

16

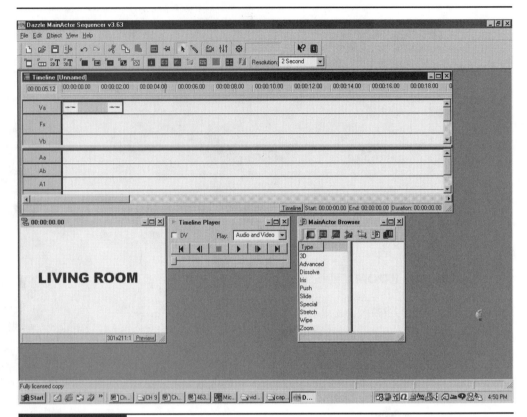

FIGURE 16-9   Drag the title clip to 3:00 seconds

Creating a Final CD-ROM of Your Inventory Video

Now you are ready to put your inventory in final form. You might choose to create an edited video clip for each room and save that clip as a free-standing video or you may choose to assemble the entire video with all the rooms into one video program. I would recommend creating a separate video file for each room for ease of handling. If you have limited hard drive space you also might need to delete the original DV capture files from your hard drive for one room before moving to the next. The MPEG files that you create when exporting the file will be much smaller than the original DV files imported from the camcorder into the computer. Following is a step-by-step process for creating a final CD-ROM of your inventory video, using a separate video file for each room.

1. When you are finished assembling and editing the first room video program, save the project for future use and then select File | Export. This will open the Export dialog box, shown here:

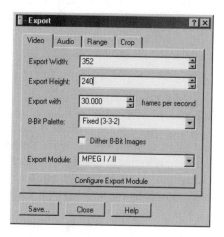

2. Change the default settings to 352 pixels wide by 240 pixels high. Select MPEG I/II from the Export Module list to export your program to MPEG video file format for creation and playback from a CD-ROM.

3. Click Save and choose a file name and a file location to store your MPEG video clip in the Select File dialog box that appears, as shown in the illustration that follows. Click Save when you are ready and the export process will begin. It may take a few minutes for the process to work depending on the length of the program in the edit timeline.

4. When you have saved your first clip, create a new timeline and assemble the next room sequence of clips and titles.

16

When all your clips are assembled, use the procedures described in Chapter 14 to create a CD-ROM that contains your video clips and any word processing or database files containing information relevant to your household inventory. Store a copy at your office, in your bank safety deposit box, or another location outside your home. I hope you never need to use the video for its intended purpose, but creating a good record of your possessions can give you peace of mind.

Summary

Although it's not something we like to think about, everyone knows someone who has had a fire or a burglary and had to go through the difficult and emotionally draining experience of constructing an exact list of possessions and assessing their value for insurance purposes. A video inventory is much more effective and helpful than still pictures, as it can show items in their locations with accompanying audio explanations. Such an inventory can also be edited, which is good film-making practice as well as a practical alternative to producing a potentially valuable resource.

Chapter 17

Make a How-to Video

How to...

- Plan a how-to video
- Create a script
- Use screen camera software to enhance your video

How-to videos are very popular today. They are all about teaching someone how to do something by demonstrating a process on tape, with a narration providing explanation and description step-by-step. You find how-to videos used in business for a variety of training purposes, and at the home video store for people who want to learn how to do a wide variety of things, from playing the guitar to making a pie to painting landscapes. The good videos all have one important thing in common—a good script. Scripts for how-to projects have to break the project into clear, concise, and accurate steps that move the viewer through the process without a hitch. It is all too easy for someone who is knowledgeable about a process to write a how-to script with holes in it, leaving out steps or concepts that seem obvious or logical to the expert, but not to the learner. When making a piecrust, the expert baker knows not to handle the dough for too long. If the script doesn't give the viewer this necessary piece of information, you have inedible piecrust and a disgruntled would-be baker who will not think highly of the teacher. I have included a fairly detailed script to use as an example of telling your how-to story as well as thinking through your production and editing process. I don't intend for you to copy it literally, just to be guided by my experience in your endeavor.

On the technical side, the project we will talk about here introduces screen camera software combined with camcorder footage and interactive video for CD-ROM and Internet. We will script and create a short segment of a how-to video demonstrating the process of using a website.

This chapter assumes that you have certain basic video skills or have read earlier chapters of the book. The skills and the chapters that cover them are as follows:

- *Using your digital camcorder (Chapter 1)*
- *Shooting quality video (Chapters 2 and 3)*
- *Video lighting (Chapter 4)*
- *Video sound (Chapter 5)*
- *Production preplanning and scripting (Chapter 6)*

Scripting Your Project

Let's look at a script created for a real project. Remember, you might need to review the more in-depth coverage of preproduction planning and script writing in Chapter 6 before you begin. DocsDepot.com is a company I work with and for whom I designed an e-business website that provides Internet-based purchasing services for physicians (see Figure 17-1). The company allowed me to use their website catalog as the subject of the how-to project for this chapter.

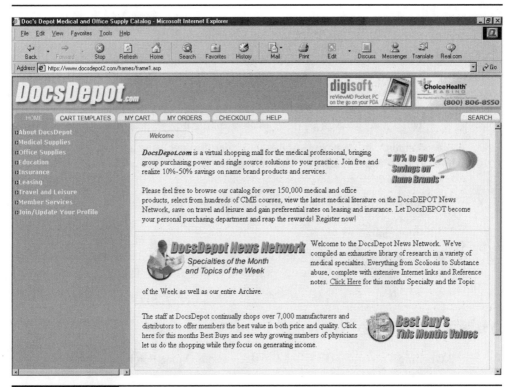

FIGURE 17-1   DocsDepot.com website

The video combines live digital video with animated video recording of a computer screen using a software package called HyperCam, a shareware product available from http://www.hyperionics.com/. You will be recording most of the voiceover for the video while running the DocsDepot.com or other application software or website and recording your demonstration using the HyperCam software.

HyperCam records the screen animation to a series of AVI digital video files that can be combined with the live DV video clips using your standard video editing software just as if it had been recorded using a video camera. There are advantages of using this software instead of pointing the camera at the screen and recording. The principal advantage is that the video monitor of a computer scans at a different rate than camcorders. The result is that there will be a periodic black line traveling from the top of the screen to the bottom ruining your video. Another problem is that it is difficult to line up the camera with the screen properly without external equipment.

I have included a complete script for the project in the next section. It can serve as an idea-generator for your own project as well as one example of a form for video scripting. The form of your script may vary, depending on the type of video you are developing and the topics required to describe it.

17

Set Up Your Equipment

The equipment you will need for this shoot may include:

- **Video camcorder and tape** This is the minimum equipment, of course. This will be needed if you shoot interviews or tours of a facility as part of your presentation. You can do an excellent job of recording your organization's video with only this basic equipment. If you are not shooting live video and only recording from the computer screen, you will not need a camcorder to create a similar how-to program.

- **Tripod** In many cases it may be easier to shoot using a camcorder mounted on a tripod. You will have to think through each shot and how it can be best captured as well as what equipment you have available.

- **Extra lighting** Depending on the shots you are planning, particularly if you do interviews, you should consider extra lighting. Refer to Chapter 4 for more detailed information.

- **HyperCam software** You will need to download and install it in order to begin recording.

Video Shooting Script

Opening Title: Help Using the DocsDepot.com Online Purchasing System

Time: 10 seconds

Description: Basic title over a texture background

Transition: Cut

Audio: Music track from stock music library

Voiceover: None

Scene 1

Time: 30 seconds

Description: Introduction by DocsDepot.com president

Camera: Head shot

Transition: Cut

Audio: Recorded with clip

Voiceover: None

Scene 2

Time: 30 seconds

Description: Browsing online catalogs

Camera: Screen cam recording

Transition: Cut

Audio: Recorded with clip

Voiceover: To locate an item, select a catalog, either Medical Supplies or Office Supplies, from the list on the left by clicking the title. When the table of contents appears, click the selection that best describes the category of the item. When a table of contents search has been done, a list of products that meet the search criterion will be displayed. On the right side of the list of products are View buttons. When these are clicked, a detailed view of the specific product including any pictures, options, and other selection information will be displayed with buttons that allow for selection and saving to your shopping cart.

Scene 3

Time: 30 seconds

Description: Searching online catalogs

Camera: Screen cam recording

Transition: Cut

Audio: Recorded with clip

Voiceover: Items may be located directly by clicking the Search button located on the right side of the header bar at the top of the screen. You may search any of the database categories listed in the Category pull-down menu. You must then select the kind of search you want: Keyword(s), Product Name, Product SKU (the stock keeping unit or catalog number used by the purchasing group), or the Manufacturer SKU (the stock keeping unit or catalog number from the original manufacturer). Some product or manufacturer SKUs (on products with options) are contained in the Keyword field. If you don't find the product you are searching for using either of the SKU fields, try the Keyword field. Then enter the keyword, product name, or SKU you wish to search for and click Search Now. The Clear button sets the search screen to default for new entries and the Done button returns you to the main screen.

Scene 4

Time: 30 seconds

Description: How to place an order

Camera: Screen cam recording

Transition: Cut

Audio: Recorded with clip

Voiceover: Each item in the product list contains an Item Quantity box and an Add To Cart button. To order a product, enter the quantity of items in the Item Quantity box. Click the Add To Cart button beneath the item to add the item to the cart. The Cart list will display the price, quantity, and extended cost of the item, as well as the total cost of the order. If the product has an option that must be selected to make a valid order, you must use the option pull-down menu or other option selection process and then click the Update Price button before you will be allowed to add the product to your shopping cart. When this is done, a red message that states that the product has been "optioned" will appear on the blue Product Name field.

Scene 5

Time: 30 seconds

Description: Using the DocsDepot.com shopping cart

Camera: Screen cam recording

Transition: Cut

Audio: Recorded with clip

Voiceover: The Cart list displays the items, quantities, and costs associated with an order. To change the quantity of an item, click the Item Quantity box and enter the new quantity. Click the Recalculate button to adjust the total cost of the item and total cost of the order. To remove an item from the cart, click the check box next to the product code for the item. Click the Recalculate button to remove the item from the cart and update the total cost of the order. To empty the cart, click the Clear Cart button. This will remove all items and recalculate the total cost to zero. Office and medical supply carts will be combined as you complete your order. Once you have added all the items you wish to purchase, click on the Check Out button to allow us to place your order.

Scene 6

Time: 30 seconds

Description: DocsDepot.com shopping cart templates

Camera: Screen cam recording

Transition: Cut

Audio: Recorded with clip

Voiceover: We have provided the ability for you to save shopping carts as templates when you are placing an order. This will allow you to easily repeat an order for common supplies and products. We have also provided a series of global templates, by medical specialty, which give you a starting place for creating orders for commonly purchased products and supplies within your practice specialty. You must enter your member ID and password to view and select your templates or global templates. If you wish to use global templates when placing your first order, you must register as a member.

Scene 7

Time: 30 seconds

Description: Checking out from the DocsDepot.com web store

Camera: Screen cam recording

Transition: Cut

Audio: Recorded with clip

Voiceover: You will be asked to log in with the membership username and password you selected upon registering. Billing information and any additional required information for your order will be confirmed after you have logged in. The Checkout screen displays the complete order including total cost. To allow substitutions for items that are out of stock or unavailable, click the Allow Subs button to the left of the item in the Allow Subs column. To change the quantities of the items listed before the order is sent, click My Cart in the header bar at the top of the screen to return to the cart. To exit the Checkout screen without ordering items, click the Cancel button. To purchase the items listed, click the Send Order button. You should receive a confirmation of your order.

17

Scene 8

Time: 30 seconds

Description: Checking your order status

Camera: Screen cam recording

Transition: Cut

Audio: Recorded with clip

Voiceover: Selecting My Orders will allow you to see a list of the orders you have placed. To review the details and status of each order you have placed, and the status of each item on the order, click the Order ID you want to view and a detailed view will be displayed.

Creating Your How-to Video Using HyperCam Screen Camera Software

1. Download the HyperCam software from http://www.hyperionics.com/.

2. Register the product for $30 (you can try it first).

3. Install it on your computer according to the instructions.

4. From your Windows Start menu, select Settings | Control Panel | Display. The Display Properties dialog box opens, as shown in Figure 17-2. Choose the Settings tab and set the screen area to 640 x 480 and at least 256 colors for capture. I usually use True Color (32 bit) for video editing. You can set it back to a larger area and more colors for your normal video edit use. It is also possible to capture from a section of a larger screen and with more colors, but it can be faster and more effective if set to 640 x 480 temporarily.

5. Select the Background tab, and set Wallpaper to None, as shown in Figure 17-3. This creates a black background, which will hide any elements that might conflict with your application screen.

6. Select the Appearance tab and set the Scheme to Windows Standard as shown in Figure 17-4. Using the pull-down menus, you should adjust the colors of the items that are blue to black, and the items that are off-white to white. You can save this scheme under a different name so that you can easily reset your normal screen and recover the recording setting for another session. Close the Display Properties dialog box.

7. From the Control Panel, select Mouse. The Mouse Properties dialog box will appear, as shown in Figure 17-5. Select the Pointers tab and set the Scheme to Windows Standard (extra large). The larger pointer will be easier to read in the edited video and eliminate the strobing effect when recording the movement of the small arrow. Close the Mouse Properties dialog box.

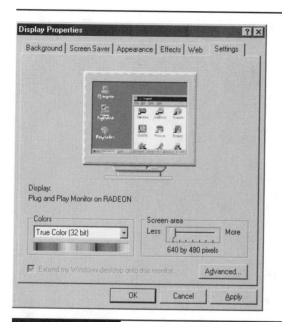

FIGURE 17-2 HyperCam display settings

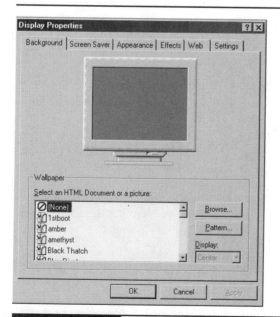

FIGURE 17-3 Background tab

17

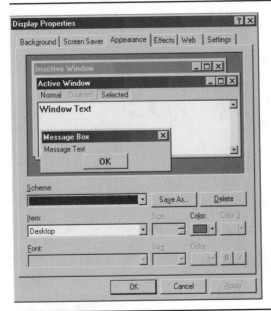

FIGURE 17-4   Appearance tab

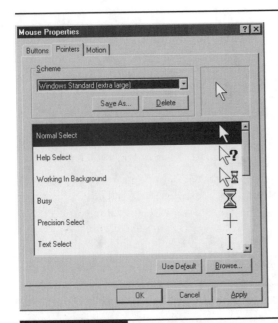

FIGURE 17-5   Mouse Properties dialog box

8. Resize the task bar at the bottom of the screen to three times as tall as it is normally. Do this by dragging the top of the task bar with your mouse until it is resized properly. When you have finished, close the Mouse Properties dialog box.

9. Click the HyperCam icon to begin the HyperCam recording session. The control panel is shown in Figure 17-6.

10. In the AVI File tab, set the HyperCam video recording to 640 x 428 at 4 or 5 frames per second.

11. In the Sound tab, set sound recording at 11025 samples per second, 16-bit. This is the minimum setting for voiceover recording.

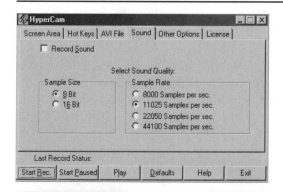

FIGURE 17-6   HyperCam control panel

 It is more important for voice recording to have the sample rate low and setting high. The quality of voice recording will not be reduced at this

12. Connect a microphone. A good directional microphone will work, bu' microphone will free your hands to move the mouse and turn script p

13. From your Start menu, select Programs | Accessories | Entertainme (you may need to install this from your Windows CD). This w' control panel for your computer. The number and type of contr'cursor. look different than Figure 17-7 as every computer manufactv'. When the sound control panel.   t back to

14. Set the level of your microphone by dragging the volume You might need to readjust the sound level if it record' you begin the recording process, you should record make sure the settings are optimal. Leave this win' check in steps 15 through 17.

of

rm a sound

17

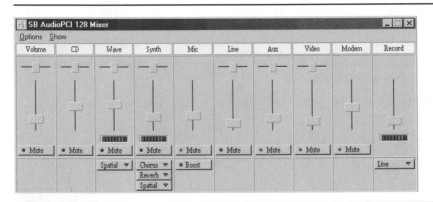

FIGURE 17-7   Volume control panel

15. From your Start menu, select Programs | Accessories | Entertainment | Sound Recorder (you may need to install this from your Windows CD). This will open the Sound Recorder utility. The control panel will look like the one shown in the illustration that follows. When you are using HyperCam, this utility is what will be recording your voiceover and other sound.

he volume control panel, set the Microphone Balance at the highest volume
stortion.

Recording Balance, on the volume control panel, should be about in the

adm control panel, make a test recording with HyperCam and make
Ma peak normally the indicator flashes green and yellow, not red (if
middl will sound distorted on playback). You may need to adjust the
line. Pla ne (try about three or four inches from your mouth) and/or
no hiss or Balance. Play the video back through headphones. Switch
ow the playback volume. Put the Wave Balance in the
ut Volume Control Balance at the second lowest setting
g—it should be at a comfortable level with little or

Pay attention to background noise, noise from your computer's hard drive, and any mouth and breathing noises you might not be aware you are making.

19. Make your final recording. Figure 17-8 shows one frame from the recording of the DocsDepot.com demo.

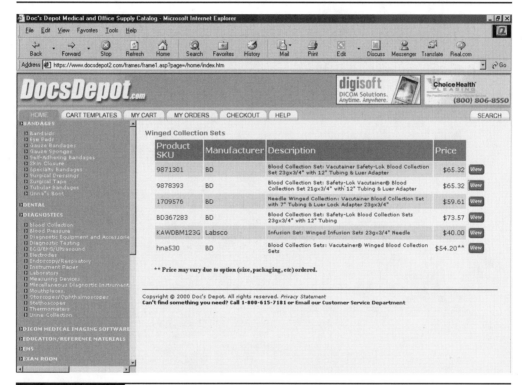

FIGURE 17-8 Video frame of DocsDepot.com demo recording

Things will go faster if you have a verbatim script that is well rehearsed. You can always make changes as they are needed. It may be easier to have one person read the script and another operate the mouse and keyboard. This will give both elements the benefit of total focus. It may take several tries at the recording to get it right. Be patient—it will show in the quality of the work.

Summary

Once you have recorded your demonstration using HyperCam, you should then edit the program using the techniques you learned in Chapters 6 through 12. How-to videos depend on clear and well-planned scripts. You must be familiar with the process you are describing if you are going to be able to describe it well to someone else. With a well-written script, your viewer should be able to follow a procedure without any missing concepts or steps. Then it's just a case of "ready, set, go!"

Chapter 18

Create a Marketing Video for Your Organization

How to...

■ Plan a video for your organization

■ Create a shooting script

■ Organize a shooting schedule

■ Record voiceover

Nearly everyone belongs to an organization of some kind. Organizations include everything from professional associations to the Girl Scouts, from civic organizations and the garden club to churches and synagogues. Most organizations have an interest in recording their history and their identity, their interests and their growth. Videos are created to celebrate organizational anniversaries, fund-raisers, membership drives, and special events. Acquiring skill as a videographer—as you are doing in this book—can be a real asset to the organizations you belong to. Creating and editing a finished and professional-looking video document can be an important contribution to any organization.

> **NOTE**   *This chapter assumes that you have certain skills or have read earlier chapters of the book. The skills and the chapters that cover them are as follows:*
>
> ■ *Using your digital camcorder (Chapter 1)*
> ■ *Shooting quality video (Chapters 2 and 3)*
> ■ *Video lighting (Chapter 4)*
> ■ *Video sound (Chapter 5)*
> ■ *Production preplanning and scripting (Chapter 6)*
> ■ *Transitions (Chapter 10)*

Regardless of the type of organization you are working with, advance planning and organization of the project will save you time and effort in the long run, and will make the project a creative and rewarding one. In the first stages of planning, you should consider such topics as:

The purpose of the video   Although many types of video productions will probably include basic information about the organization, the level of detail will vary and the focus will be different from project to project. Firmly determine what aspects of the group you want to highlight and design your project accordingly. This is called a *terminal objective*. As a rule, people usually retain a maximum of three main points from presentations of information and data. Keep in mind, as you write your script, this rule of thumb: after viewing this program, the audience will understand no more than three main concepts. Begin your script writing with a sentence that states clearly and concisely the main points you want to convey in the video. This cannot be stressed too much. Once established, this sentence should guide everything that follows in the scripting process. As you write the script, you should keep in mind the objective in every sentence you write. If it does not support these objectives, get rid of it.

The length of the video   Length is an important determining factor in creating a video. Typically, a marketing video is short and snappy, while a historical document is longer and has more detail. No one video can do everything, so be discriminating in planning content so it will fit the length of viewing time you are working with. Even in longer projects, it is best to say what you have to say without too much extra material. In video projects, less is frequently more. Naturally, as with all video projects, you should *shoot* extra material so you have plenty to choose from, but be selective in the editing process so your video is crisp, to the point, and does not belabor the subject to the yawning point. Either a video is too short, leaving the audience wanting more, or it is too long, and they turn it off. Too short can motivate them to follow up on their own. Too long and you have lost your audience for good.

The audience for the video   A video directed at young adults may have an entirely different feel than one aimed at seniors or kids. Music, for example, is an excellent element in a video project, but the music in a video whose audience is primarily seniors will obviously differ from that aimed at teenagers—or it should!

Creating a Marketing Video for an Organization

Today, all types of organizations are interested in creating videos, both as historical documents and as marketing tools. Whether the organization is a civic group promoting local businesses and cultural events, a group devoted to the arts, a political organization wanting to spread their message, or a church, synagogue, or mosque interested in increasing membership, a marketing video is an excellent and increasingly popular way of getting the word out. For a civic organization, you might want to make short marketing videos available at the Chamber of Commerce, the local library, or for sale at a modest fee to new people in town, as well as merchants and others participating in the organization. Many religious organizations are creating marketing videos, and such video projects can be valuable aids in conveying the basic identity and character of a church, giving visitors a feel for the programs and the people who worship there. This chapter will use a church video project as an example of the kind of dynamic documentation a video can provide for any organization.

In planning such a video project, it is important to have as comprehensive an overview of the goal of the video as possible. There are important questions to be asked before you begin shooting, beginning with the principal purpose of the video. Is it to document your church and its programs for historical purposes, or is it a marketing tool? Is the video being created primarily for church members and their library, or is it for guests to acquaint them with the church and its members? The answers will give you an important head start in planning your video project. This example will create a marketing video for a church seeking membership growth, but the same basic approach works for most organizational projects.

Regardless of the organization being represented, marketing videos are most effective when they are short and contain vivid images that are expressive of your message. In the case of the proposed church project, the video will be eight minutes, forty seconds long, and will be given to visitors in their "welcome packets" at the morning worship service. This is an important strategy as it enables visitors to take the video home to watch. The video will answer a lot of questions about the church, and reinforce the impression made during their visit. The primary consideration during the planning process should be what images will convey your message to viewers. The following questions focus on the church video project, but many of them would translate easily to other types of organizations. This will just give you ideas for how to write a script and what the script's content could be.

- What are our church's strengths? Outreach programs helping in the community? Evangelism efforts? Creative worship services? Narrow your focus to the areas of strength you want to highlight in your video.

- What people in the church are most representative of our program and character?

- What events will capture the most essential character of the church program and people? Family night dinners? Worship services? Service projects in the church or in the community?

As this is a video to be handed out to the public, it must be well made and professional. This is not to scare you away from such a project! With the knowledge and skill you have developed, you can do it and do it well. You will want to consider using titles, narration, and/or music to accompany the images. To accomplish all this in a well-organized fashion, begin with a shooting script.

Set Up Your Equipment

The equipment you will need for this shoot may include:

Video camcorder and tape   This is the minimum equipment, of course. You can do an excellent job of recording your organization's video with only this basic equipment.

Tripod   In many cases, it may be easier to shoot using a camcorder mounted on a tripod. You will have to think through each shot and how it can be best captured as well as what equipment you have available.

Extra lighting   Depending on the shot you are planning, particularly if you do interviews, you should consider extra lighting. Refer to Chapter 4 for more detailed information.

Video Shooting Script

Title 1: Opening Title

Title text: Vallejo First Christian Church: A Church for Two Millennia (see Figure 18-1)

Time: 20 seconds

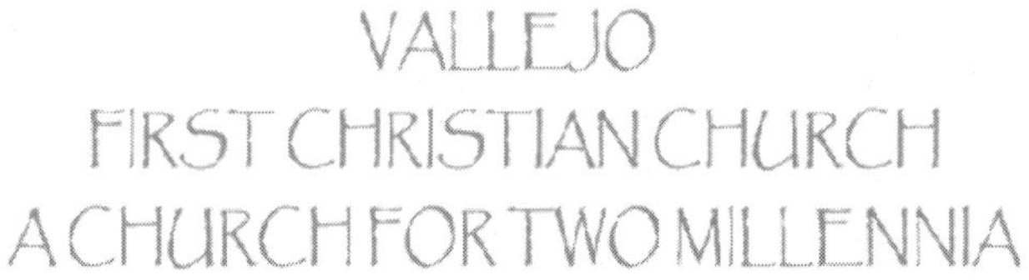

VALLEJO
FIRST CHRISTIAN CHURCH
A CHURCH FOR TWO MILLENNIA

FIGURE 18-1   Title 1

Description: Display this title slide over a photo of current church, including rainbow flag

Transition: Cut

Audio: Trio playing/singing a hymn

Voiceover: None

Title 2: Second Title

Title text: Vallejo First Christian Church (Disciples of Christ), est. 1899

Description: Video of people entering the church with title superimposed

Time: 30 seconds

Transition: Cut

Audio: Trio instrumental, playing a hymn under voiceover

Voiceover: The voiceover for this part of the script could describe some brief historical highlights including the date of the founding of the church, the number of original members, and the mission that provided impetus for building the church.

Scene 1: Construction of the Building in 1943

Time: 30 seconds

Description: Picture sequence of construction of the building from 1943 to present, using shots taken from the same angle of the building (see Figure 18-2)

Picture 1: Frame construction of building, taken from south

Picture 2: Scaffolding around building, taken from south

Picture 3: Finished building, taken from south

Picture 4: Building today, taken from south

Camera: N/A

Transition: Cross fade from pictures 1–3, hold on picture 4

Audio: Trio instrumental, hymn under voiceover

Voiceover: The script could describe some details about the building currently used by the church, including dates of construction and any special circumstances regarding the building process.

FIGURE 18-2   Scene 1

Scene 2: The Building Today

Time: 15 seconds

Description: Video of outside of building shot from the south, same angle as previous photos. Shot to include rainbow flag and people gathered outside around produce table following Sunday service (see Figure 18-3)

FIGURE 18-3   Scene 2

Camera: South of church, locked down

> Shot 1: South side/front of church including sign with rainbow flag

> Shot 2: Zoom on rainbow flag

Transition: Cross fade from previous picture sequence; zoom to close-up of flag

Audio: Trio instrumental playing a hymn under voiceover

Voiceover: The script could describe the church as it is today, with a brief mission statement.

Scene 3: A Lively Tradition

Time: 30 seconds

Description: Video of people gathered outside church following worship (see Figure 18-4)

Camera: Panning left to right facing the church from the east

> Shot 1: Front of church

> Shot 2: Zoom in on group

> Shot 3: Close-up of one of the group

Shot 4: General shot of people talking

Transition: Cut from previous video

Audio: Ambient sound of people talking

Voiceover: The script could describe various outreach or mission projects in which the church participates.

Scene 3

Scene 4: A Space for Worship—Then and Now

Time: 15 seconds

Description: Montage of photos of sanctuary (see Figure 18-5)

Photo 1: Sanctuary scaffolding

Photo 2: Sanctuary with pews

Photo 3: Sanctuary with Pentecost banners

Camera: N/A

FIGURE 18-5   Scene 4

Transition: Cross fade from pictures 1–3, hold on picture 3

Audio: Organ playing appropriate prelude

Voiceover: The script could describe the interior of the church sanctuary, and talk about the types of worship that occur there.

Scene 5: Recent Art Installations

Time: 15 seconds

Description: Montage of photos of art installations in the sanctuary (see Figure 18-6)

> Photo 1: Pentecost
>
> Photo 2: Installation service
>
> Photo 3: Easter banner

Camera: N/A

Transition: Cross fade from picture 3, scene 4, and between pictures 1, 2, and 3, scene 5

Audio: Organ playing appropriate prelude

Voiceover: The script could include descriptions of art or banners that are displayed in the sanctuary for various seasonal events.

FIGURE 18-6   Scene 5

Scene 6: Communion

Time: 30 seconds

Description: Video of congregation sharing in common cup communion service (see Figure 18-7)

    Shot 1: People lining up for communion

    Shot 2: Zoom in for close-up of faces

    Shot 3: Close-up of bread

    Shot 4: Close-up of wine

FIGURE 18-7   Scene 6

Camera:

> Shots 1 & 2: Shot from chancel of people lining up in aisle; zoom in on close-up of faces

> Shots 3 & 4: Camera to the front/side of celebrants for close-up of bread; close-up of cup

Transition: Cut

Audio: Organ music, "One Bread, One Body"

Voiceover: The script could describe a special service celebrated in the sanctuary, or regular elements of Sunday worship.

Scene 7: Preaching

Time: 15 seconds

Description: Shot from an angle/front of sanctuary of pulpit and pastor preaching (see Figure 18-8)

FIGURE 18-8   Scene 7

Shot 1: Pulpit and pastor preaching

Shot 2: Close-up of pastor

Camera: Locked down; zoom in on close-up of face

Transition: Cut

Audio: Clip of preaching following voiceover

Voiceover: The script could describe the specific ministry of the pastor of the congregation.

Scene 8: Choir

Time: 20 seconds

Description: Video shot from center of sanctuary of choir singing (see Figure 18-9)

Camera: Locked down

Transition: Cut

Audio: Choir singing

Voiceover: None

FIGURE 18-9   Scene 8

Scene 9: Guitar Soloist

Time: 20 seconds

Description: Video of guitarist (see Figure 18-10)

   Shot 1: From sanctuary, medium shot, face-on of guitarist playing and singing

   Shot 2: Medium close shot, side view of guitarist

   Shot 3: Close-up of hands and face

Camera: Locked down from the front; locked down from the side, zoom onto hands and face of guitarist

Transition: Cut from previous video; cross fade between shots

Audio: Guitar and singing

Voiceover: None

FIGURE 18-10   Scene 9

Scene 10: Special Music

Time: 20 seconds

Description: Video of musicians playing guitar and bass and singing (see Figure 18-11)

    Shot 1: From right-center sanctuary

    Shot 2: From center-front sanctuary, side view

    Shot 3: Close-up of hands of bass player

    Shot 4: Close-up of singer

Camera: Locked down from left front, diagonal view of musicians; right-center sanctuary; center-front sanctuary; zoom in for close-ups

Transition: Cut from previous video

Audio: Group singing

Voiceover: The script could describe the types of music and musicians that contribute to the worship program of the church.

FIGURE 18-11   Scene 10

Scene 11: Kick-It Klub

Time: 30 seconds

Description: Kick-It-Klub kids (see Figure 18-12)

    Shot 1: Kids outside fellowship hall

    Shot 2: Video of kids assembling health kits

Camera: Slow pan; close-up of hands and faces

Transition: Cut from previous video

Audio: Ambient sound of kids talking

Voiceover: The script describes the programs that involve children in the life of the church.

FIGURE 18-12   Scene 11

Scene 12: Who Are We?

Time: 45 seconds

Description: Montage of film clips of members of all ages (see Figure 18-13)

    Shot 1: Five-year-old running up the aisle

    Shot 2: 90-year-old at his birthday party

    Shot 3: Elderly man leaving church

    Shot 4: Teenagers serving communion

    Shot 5: Mother with two kids

    Shot 6: Middle-aged married couple

    Shot 7: Big group of members on church steps

Camera: N/A

FIGURE 18-13   Scene 12

Transition: Cut from previous clip; cross-fade between shots

Audio: Choir singing an anthem

Voiceover: The script here could point out the nature and diversity of the members of the church, and recap the mission statement.

Scene 13: Rainbow Coalition

Time: 30 seconds

Description: Close-up of rainbow flag (see Figure 18-14)

Camera: Locked down from front left of flag; fade out

Transition: Cut from previous shot

Audio: Choir singing hymn

Voiceover: The script should conclude with a closing statement of welcome and invitation.

FIGURE 18-14   Scene 13

Title 3: Credits

Time: 30 seconds

Description: A series of slides with the names, contributions of the participants (see Figure 18-15)

Camera: N/A

Transition: N/A

Audio: Choir singing

Voiceover: None

Total run-time: 8:40 minutes

FIGURE 18-15   Title 3

Shooting Your Scenes

Many of the scenes in the previous script are taken from video archives that were shot in the year prior to making the program. If your organization is considering an ongoing marketing video, you might consider planning to shoot special events and other scenes to use for periodical updates to your marketing video.

TIP
Purchase a FireWire external hard drive to store your edit files from your ongoing marketing video and use it as backup and archive for the project. Then you can simply import and add or exchange more recent scenes to your video and make an updated version for marketing purposes. You can purchase a fast 30 to 60 MB FireWire hard drive for $200–300.

You should be prepared for a variety of scene types for a project like this. Pay attention to audio and lighting continuity when you shoot so the scenes will not change color and ambient sound in a distracting manner. Refer to Chapters 1 through 4 for more detail on using your camcorder and shooting good video.

Scanning Still Images

A number of the scenes in the church project are created from still images. Some of the images were captured using a still digital camera. These were JPEG images shot at 640 x 480 resolution, and the images were imported directly with no modification into the video timeline to create the still image animation. Older images were scanned using a flatbed scanner. They were also scanned at 640 x 480 resolution in 16 million colors and saved as highest quality JPEG images. Some of the images needed to be cropped and resized using Photoshop before they were imported into the video edit timeline. JPEG images are very compatible with video clips and require little or no processing.

Designing Your Sound

The soundtrack for the church project consists of three elements: the sound captured when some of the video clips were shot, voiceover recorded during the editing process, and music tracks. Care was taken to maintain high quality sound recording with video, and everything was shot with external microphones. Chapter 5 covers sound recording, and Chapter 11 covers sound editing and processing with your computer.

Making Music Selections

The music selected for the example used in this chapter was either classic church hymns or original works composed by church members. The music was recorded directly to the digital video camcorder using cables from the church sound system and audio mixer board. The music track was later extracted from the video on the computer and added to the timeline as required for a musical score. For other types of organizations, you would use music that captures the feeling and character of the individual group. A political organization might use patriotic music recorded by a local singing group or obtained through licensing agreements. A wonderful environmental organization's video used an old recording of Pete Seeger's "Goofing Off Suite" for banjo. A video for Greenpeace might use recorded whale song. Let your imagination loose! Chapter 5 covers sound in more detail.

You can improve the quality of your external sound connection if you add a balanced audio input adapter to your camcorder. These are generally available as accessories from your camcorder manufacturer or from after-market companies. They consist of a black box with professional XLR microphone connectors and have wires to plug into the camcorder's audio input jacks.

Recording Voiceover

In the previous video project, voiceover was recorded directly to the computer using a desktop microphone and Sound Forge as an audio recording and editing software package. The voiceover clips were then imported into the edit timeline and placed in the proper locations. You may find this an easy way to incorporate voiceover in your video project. Refer to Chapter 5 for more detailed instructions in recording sound and voiceover.

Make sure your microphone is well away from the computer itself. Microphones easily pick up hard drive noise. We become so used to the background noise that we can easily overlook it until we hear it in the background of recordings. Make a test and listen carefully.

Checking Your Copyrights

Make sure that you don't select music that is in copyright without clearing the copyright with the copyright owner. This is usually too expensive to do for amateur productions. That is why I chose to use classic hymns that were in the public domain or out of copyright, or to use original compositions by church members who would grant permission without cost. See Chapter 13 for more information on copyrights and check with your attorney for further information regarding your situation.

Editing Your Clips

Once you have assembled the still images, videotapes, and audio recordings for your project, you should convert the shooting script into an editing script. Tag all the clips and scenes, and begin the process of transferring your audio and video into the computer. Carefully assemble the clips into your timeline and experiment with timing and pacing. The most difficult thing you will face will be creating a steady pace for your project without it becoming tedious and boring. Use the music as a tempo guide and select clips and adjust the animation of still image montages to move as if they were dancers moving to the music track. Chapters 6 through 12 cover the editing process.

Creating Your Titles

The titles should be designed appropriately for your production. They should let the audience know what is coming, who is presenting it, and why they should be interested. Don't scrimp on the credits. This is an opportunity to build goodwill with your associates and set the stage for future projects. Check out Chapter 12 for more detail on title design and production.

Use some of your still images as an animated montage background during your title sequence.

Transferring Your Finished Video Out to Tape

Create a clean preview of your video for review. Make any last minute adjustments. Make at least two master copies of the project on digital tape, one for backup and one to use as the master for duplicating. See Chapter 13 for more information about duplication and CD-ROM production.

Use professional duplicating tapes with only enough tape to record your production (10, 15, or 30 minute tapes).

Summary

This design for a church promotional video will give you an idea of how such an organizational video can be constructed. This video attempts to include the many different aspects of the organization, a sense of the people who belong, and reasons why a prospective member might want to join. These elements apply to most organizations, and the outline of the video can be adapted to fit. Perhaps the most important part of constructing such a video project is the planning process during which you come to a clear understanding of the purpose of the video. No one video project can cover all aspects of any organization, and as you want the video to be short and snappy, you must plan carefully to make sure it makes the points you want to make quickly and clearly. Look at all the elements as creating a harmonious whole—the music track, the style of language used in the voiceover, and the images. All the elements should come together to make the statement you want to make about your organization.

How to...

- Make a living family history
- Make the perfect wedding video
- Create an on-line video course

We've done quite a lot together at this point, moving through the processes of operating a digital video camera, designing, organizing, shooting, and editing video projects. From this point on, practice and your creativity will open the fascinating world of digital video to you in ways you may never have thought of. This chapter is a grab bag of ideas for projects. Dip in and pull one out, or let the ideas here spark new and original ideas. Either way, ready, set, shoot video!

Making a Living Family History

There has been a real explosion of interest in researching family history in America. People search for their roots in national archives, through various genealogical societies, and online through the numerous genealogy search programs available. A video project including interviews, family homes and locations, and photographs will be not only interesting for your family, it will be a historical document. A real family history is not one single videotape, or two, but a library of tapes with interviews of as many people as you want to include. While compiling such a library of tapes can be time-consuming, once you have your basic script planned, you can take advantage of family get-togethers to tape one or two people. Start with your oldest living relatives. Each videotape will be a record of the family from one person's perspective. A library of such tapes will be an enormously valuable family history that you and your family will treasure.

You can also make a single-family history tape with clips from various family members, making use of the best stories from each. This will be a very popular and useful tape for sharing with family. Be sure to create titles identifying each person interviewed. Before you can make the short version, however, you have to gather all those stories that will make up your living family history library.

Interviews

A video of family members reminiscing about their lives can be an irreplaceable record of your family's history. Some years ago I started videotaping members of my family, encouraging them to tell me about their earliest memories, sharing their life's journey, and inviting them to give personal words of wisdom to the younger generations. With sadness I think of the grandparents, aunts, and uncles who were gone before the availability of video cameras made this kind of project feasible. We have some written records, but seeing the relatives on tape, hearing their voices, seeing their expressions, and sharing their laughter is an opportunity to know them in a way the written record does not allow. There are people in your family who can share fascinating memories of the past with generations to come. Seek out the storytellers in your family, the ones who remember the old days, the ones who love to sit after dinner and reminisce about their youth. You might also consider starting a videotape library of family members interviewed at different ages, so you have material to draw on in the future.

A good interview requires some advance planning and organization, like all good video projects. Some family members will show reluctance to be taped, but prevail upon them as the storehouses of memory they are. Enlist their cooperation by letting them know how important you consider their contribution to the family history. Without doubt, the most important thing about this kind of interview, the thing that will ensure success, is a well-planned script. People will ramble, of course, but it's always better to have more material than you need. A good script will prevent you from forgetting some very important questions—and it is not always possible to go back and ask them later. Write out the script and be prepared to encourage your family to relax and tell their stories for the children of the future.

Tell Me the Story: Sample Questions

While these sample questions here will provoke a lot of detail from your subjects, the most important interviewing skill you can develop is the ability to listen and ask follow-up questions. Don't get so tied up in your question list that you breeze by an opportunity to follow up on a story line you may never have considered. Also pay attention to the emotions of your subject while interviewing. Often you will hit a "hot button" that really is important to your subject. When you come across these hot buttons, pursue them. The best interviews are not just the retelling of facts, but the emotion that comes along with the telling.

Start with these sample questions, and don't be afraid to add your own to this list:

1. When and where were you born?

2. Tell me about the house you were born in—what do you remember about it?

    Encourage fairly detailed descriptions. Most people remember one house of their childhood as being "home." It may not be the house they were born in, but talking about the family home can bring out memories of great stories. Find out if the house is still standing, and where it is.

3. What was the town like?

4. What did your family consist of when you were growing up?

    Encourage the person being interviewed to describe family members and relationships. I discovered that my grandmother lived with her mother when my mother was small. My great-grandmother took care of my mother and her brother and sister, while my grandmother rode a streetcar from Commerce, Oklahoma, to Joplin, Missouri, seven days a week for the queenly sum of $1.00 a day for working in a laundry. My grandmother's youngest sister ran the house as a boarding house for miners, rising before dawn to cook and clean. These relationships led to a plethora of fascinating stories about growing up in a small mining town in eastern Oklahoma.

5. Do you know when your family first came to America, and where they were from?

No one in my family knows this for sure. I met a man once on a plane who had the same spelling as my family name, not a common one. We commented on this, and he said that his ancestor owned one of the ships the original Pilgrims used to come to America. From then on, we joyfully exclaimed, "We did not come over on the Mayflower. We owned it!" Meanwhile, we are doing research to find out when we really came over and how. I suspect it won't make quite such a good story, however.

6. What do you remember about your grandparents? Do you remember any stories about them?

I grew up listening to family stories and remember clearly when I was quite small hearing the tale of my great-grandmother, whose two little boys were kidnapped and sold to the Cherokee in the Oklahoma Territory during the 1800s. I subsequently asked every living member of that side of the family what they knew or had heard about this event. It makes quite a tale. There will be stories in your family too, if you have patience to find them.

7. What did your father (or mother) do for a living?

This can lead to a variety of other questions. For example, if Grandmother grew up on a farm, you can ask what farm life was like, how often they got to town, what kind of chores did she have, and so on. If Uncle Joe's dad was a lawyer, ask about legal issues he remembers, important or interesting cases. If Cousin Mary was a nurse, ask about hospitals when she was young and what is different now. I discovered that my grandfather was a miner who died when he was thirty-one years old from the "black lung," as silicosis was called. Of my grandmother's twelve brothers, ten died of this miner's disease before they were forty.

8. Who took care of you when you were small?

My great-grandmother—the one whose children were kidnapped and sold—took care of my mother when she was small. My mother told me that when anyone did anything wrong, if they did not instantly confess, she would "whip" all three kids, so as to be certain of getting the guilty one. People had hard lives then, even kids.

9. What were family holidays like? (Christmas, Thanksgiving, Hanukkah, and so on)

10. Did you have any family traditions about the way you celebrated holidays?

My mother remembers the children receiving one present on Christmas. She remembered one Christmas in particular, she got a book, my aunt a doll, and my uncle a red wagon. They each got a stocking with an orange in it—the only time, except for the occasional birthday parties of affluent friends—they ever saw oranges. I especially remember her telling me how grateful they were for those presents, when times were so hard and everyone had to work such long hours to provide them.

11.   What did your family do for fun?

Vacation stories can be great. Some families have strong traditions of going fishing together, or reading together in the evenings.

12.   What was school like?

My mother remembered the day when her youngest sister got into trouble for eating a banana in class. It was a one-room school, and my mother, her sister, and her older brother were all there. My aunt was told to stand in front of the class and finish her banana. Sobbing, she stood there, humiliated. Finally my uncle could bear no more. He stood up and pointed at my aunt and thundered, "Lorraine, take your seat!" She happily complied. My uncle was caned for insolence. That evening, one of my great-uncles had a talk with the teacher on his way home. No one was ever caned in that school again.

13.   Who was your best friend? What did you and your friends do for fun?

My dad terrified my sister and me when we were young, describing how he and his friends used to swing on a rope out over the river near home. One day the current got him, and he was swept down river to fetch up, most fortunately, on a partially submerged log. My dad and I visited that tiny town some years ago. I was amused to see there was a rope affixed to a tree overhanging that river. Kids never change.

14.   How did you meet your husband/wife/partner?

A friend told me that her father, a teacher incensed by the internment of the Japanese during World War II, volunteered to teach school for these Japanese-Americans. A young woman in Missouri went into the post office one day and saw a poster advertising the need for teachers in the internment camps. She was horrified, and immediately set out for California to offer her services. She and her future husband met there. My friend remembered her mother telling her about their moonlight strolls around the barbed wire compound.

15.   When and where were your children born?

One of my favorite stories came from a family friend whose grandfather was born in 1892. He was born prematurely and was so small the doctors gave him no chance. His mother refused to give up. She placed the tiny baby in a shoebox. Her wood stove had a warming oven, a compartment to the side of the stove that was used to keep things warm. This snug warm place became the baby's incubator. He grew up and lived to be 102. He loved telling people that he was "raised in a stove."

16.   What did you do for a living?

17.   Where have you lived?

18. What do you remember about historic events?

For example, if they were alive during World War II, or the Great Depression, what do they remember about life then? Where were they when President Kennedy was assassinated? Where were they on September 11th, 2001?

19. What people have been an important influence in your life?

20. What have been the great joys in your life? What have been the great sorrows?

21. What would you like to tell future generations in your family?

These questions and examples are meant to give you an idea of how to encourage story telling. Interviews with family members are the most important part of a visual family history project, so be sure to take your time, be patient, be willing to spend lots of time asking and listening, and buy plenty of tape!

As important as a good script is, it should not act as a straitjacket. Listen carefully to your subject's responses, so you can follow up on comments that lead you in an unplanned direction. Some of the best stories are elicited by such digressions. In listening to your subject, pay attention to their emotional responses. Such responses can be clues of important stories with much more feeling and interest than a simple recitation of facts.

For elderly relatives especially, encourage the person being interviewed to get comfortable. Seat them in their favorite chair, and sit so they will look at you and not at the camera. Eventually even the most nervous will forget about the camera and just tell the story.

Family Photos

Family pictures can be an important additional source for your family video. Collect and scan pictures to edit into your video. When a family member reminisces about their grandparents, for example, you can cross-fade in a photo of them during the editing process. So while Aunt Sarah recalls life with her grandfather, the viewer can see just what Grandfather looked like. Photos of people as well as houses and other important locations can help bring your family history alive. Using a copy stand can help in this process. A copy stand is a device that holds the camera in a stationary position above a flat surface, with a bright light shining down on the photo to be copied. You can add interest to the photos with zooming and other effects.

Houses and Locations

Where possible, shoot video of family homes and the places your ancestors lived. If the houses no longer exist, often pictures do, and you can scan and use them in the video in the same way you use photos of people. People love seeing where their parents and grandparents grew up. It can give a real sense of understanding to their stories.

Letters

Family letters can be a fascinating addition to your video. Many families have letters from ancestors. Have a family member with a good voice, who is the same sex and approximate age as the relative who wrote the letter, read excerpts as a voiceover to video or pictures of the author, their home, or the part of the country they lived in.

Editing It All Together

Once you have all your tape and photographs together, plan your editing script as you did in earlier chapters. Plan your transitions from videotape to photographs, and design your titles. Be sure to create an opening title that gives the name and birth date of the person being interviewed. You may want to include music or voiceover narration for parts of the videotape. All together, this can be a beautiful project, well worth the time spent making it.

Weddings

Wedding videos are very popular, and not everyone wants to pay what professional videographers get for making one. Once you become proficient with your digital video camera, it is not a question of whether you will be asked to make a wedding video, but of when. Making one can be fun, especially if you plan ahead and are prepared for what happens. Remember the maxim wedding planners and clergy all impress upon people who are getting married: there is no such thing as a perfect wedding. Everything depends on how well you roll with the inevitable glitches. As the videographer, however, you can avoid adding to the tensions (and make no mistake, weddings are very tense events) by being well prepared. A word of warning: weddings are emotional events, even the simplest ones. It takes courage to agree to make a wedding video, and many people are better off if they hire a professional. If you are willing to venture on, however, the following tips may help you survive the minefields with friendships intact.

Planning the Perfect Wedding Video

There are some very important questions to ask when you are prevailed upon to make a wedding video.

- Where is the wedding to be held? Is it inside a church, hotel, or home, or outside in a garden?

- If the wedding is to be held inside, what kind of lighting exists, and how much lighting will you need to provide? Churches can be dark, as can homes, especially if the wedding is in the evening. Find out well in advance what the place looks like at the time of day the wedding will be held, so you can plan how to light the event. If you have to shoot with available light, and often that will be the case, try your best to shoot from an angle that avoids pointing the camera into the sun or bright lights. If possible, aim for backlight. Most important, try to get light on the bride and groom's faces.

■ For outside weddings, is there a contingency plan if the weather is bad? This is very, very important for people planning either an outside wedding or reception. If you show up with only a camera and no lights for indoor shooting because you think the wedding is to be in a sunlit garden, and the wedding moves indoors due to rain— instantly you have blown the perfect wedding video. Be prepared for change.

> **TIP** *When using panning or zooming, be careful not to overdo. As always when doing either, be sure you get six to twelve seconds of the scene before and after the pan or zoom, in case they are shaky or don't work out for some reason. That way you won't miss any important scenes completely.*

■ Where will you stand during the ceremony? Many clergy are uncomfortable with cameras too close to the wedding party during the ceremony. You may have to place the camera to the side and out of people's line of vision so as not to be disruptive. This is an occasion when the zoom feature should provide close-ups, rather than trying to get too close yourself. I saw one wedding photographer recently crouching in the aisle shooting tape madly as the bride advanced down the aisle on her father's arm. No one was pleased, least of all the bride. The perfect wedding is not a place for paparazzi technique. Find a place that provides good vision and is not obtrusive. This may take some planning, so do all this in advance. The wedding is *not* the time to be finding a place to stand. If you must move during the wedding ceremony, do so quietly and discreetly. Again, nothing should distract the wedding party or the people gathered for the occasion from the center of attention—the wedding ceremony itself.

> **TIP** *If it is a religious wedding, be sure to speak to the officiating clergy ahead of time as some restrict or do not allow videotaping or photography during weddings.*

■ How much videotape will you need? Bring plenty of videotape, more than you will possibly need. It is always better to have too much than to run out.

■ What different locations will you be taping in and where are they? If they want video shot before the ceremony, of the bride arriving or at her home or hotel getting ready, find out where that will take place and leave plenty of time between locations. Making the bride wait while you find your way from the hotel to the wedding chapel is not a good idea.

TIP *Carry a roll or two of gaffers' tape (available at your photo supply or video supply store), or duct tape to tape down wires where there is a possibility people might trip over them, or to tape a microphone to a flower stand as a last minute improvisation.*

■ Where is sound important? There are some scenes where sound is very important, if you are recording the vows, for example, or interviews of guests. For these you will need to make prior arrangements for wireless microphones or hookups to the sound system being used, so you get the best possible sound. In other places, sound is not so important, and you can use music in your soundtrack.

TIP *I hate to remember all the times I was three feet short of having enough microphone cable. Plan ahead and plan for extra.*

Make a complete list of the scenes the bride and groom want included in the video. Include scenes before the wedding, during the ceremony, at the reception, and so on. Note where these scenes will take place, but be ready for sudden changes. The bride may think she wants to throw her bouquet from the curb, but may change her mind or be prevailed upon by her mother to throw it from the porch or church steps. Watch for these kinds of moves so you don't miss anything. If possible, get a rough estimate of how long these various things will take, especially if it is your first wedding. That way you won't run out of tape in the middle of a shot, thus ruining the perfect wedding video.

Scenes from the Perfect Wedding

Traditions vary, and weddings are increasingly diverse in style and content. Following is a list of typical scenes the bride and groom may want in their wedding video. Choose the ones that fit the particular wedding you're filming.

 Keep your eye open for amusing or touching scenes, as these can add interest to otherwise same-old, same-old wedding video. Of course, avoid anything that will be embarrassing or upsetting later.

■ The bride getting dressed. This has become increasingly popular lately, so be prepared to be asked. Be familiar, if possible, with the room in which the bride will dress. Find a reasonably good spot to stand and stay out of the way. Grooms almost never want to be taped getting dressed, so you probably will not be asked, but you never know.

- Guests arriving at the wedding.

- The wedding party arriving at the wedding.

- An establishing shot of the inside of the church or hall where the wedding will take place, including flowers, guests, and so on. Often these are shot from the back of the room or wedding area as it is awkward to have videographers standing up in the front before the wedding. If there is room to stand at the front and to the side, this can be a good location to capture the expressions of the guests.

- The entrance of the groom and his attendants with the person officiating at the wedding.

- The procession of the bride's attendants and the bride. This is one of the most important parts of the wedding, so be sure to find a good place to stand in advance. Pan slowly to follow them down the aisle, particularly the bride.

- The ceremony. Wedding ceremonies are very difficult to tape because the bride and the groom generally have their backs to the guests during most of it. If you are expected to record the sound of the vows, you will have to have a microphone nearby as you will probably not be able to stand close enough for your video camera's internal microphone to do a good job. The best place to stand, if it is possible, is at the front and to the side, for example, at the far end of the first row of chairs or pew. People who officiate at weddings are getting better about turning the bride and groom outward so the guests can see their faces, at least in profile, during the ceremony. If you can only see one of them, opt for the bride. The groom will thank you.

The ceremony may take place all on one level, or, if it is in a church, the bride and groom may move up to a higher level of the chancel. Be prepared to follow their movements with the camera, but you will probably not be able to follow them yourself without being disruptive.

Jewish weddings usually involve a canopy over the wedding party. This may present special problems for your line of vision, so be sure to plan where you are going to be.

> **TIP** *Attend the rehearsal if possible. This will give you the best preparation for where people will be standing and in what directions.*

- There may be soloists or special musicians during the ceremony. Arrange your microphones so you get good sound as well as a good view.

- The recessional. The bride and groom go down the aisle together, followed by their attendants. If you can be over to the side of the room so you have a good view of the processional, excellent. It can make a nice shot if you can be at the back of the room so you have a good shot of the bride and groom as they come up the aisle, but do not get in their way.

Once the wedding ceremony is over, the whole thing relaxes tremendously. People will no longer fuss at you for being in the way, and you can maneuver for the best (preplanned) shots. You may—in fact, you will—have to ask people to move who are occupying your preplanned locations. They will probably be understanding about it.

■ The throwing of the bouquet. Bridesmaids and single women guests cluster to catch the bride's bouquet, which is usually thrown over her shoulder. This may occur outside the place the wedding was held or at the reception.

■ The throwing of the bride's garter. Often there is an opportunity to tape the groom removing the garter the bride has worn for the ceremony, before he throws it over his shoulder to his groomsmen and other single men present. This may occur outside the place the wedding was held or at the reception.

After the bride and groom have left for the reception, whether it is in the same building or somewhere else, hurry so you get there in time to get pictures of them in the receiving line.

■ The receiving line. Formal weddings have receiving lines that include the bride and groom, their attendants, and their parents or other close relatives. Just a brief amount of tape will suffice, usually of someone kissing the bride or groom.

■ The bride's first dance. Usually the bride dances the first two dances with her father and her new husband. Some tape of these dances will be nice, followed by the guests dancing.

> TIP   *If there are children present, people often make them dance at weddings. It can be cute.*

■ The bride and groom cutting and feeding each other the first pieces of cake.

■ Toasts. If champagne is served, there will probably be at least one and perhaps several toasts made by the best man, the bride's father, the groom's father, and so on. They may be lengthy or short, articulate or mumbled, funny or boring. Try to be close enough to catch what is said. You can edit later.

■ Gifts. It is not considered good form to open gifts at the reception, but people do it anyway. Get a shot of the presents before they are opened, and a few of the bride and groom opening gifts and exclaiming over them. Don't overdo this as it does not really make for interesting viewing.

■ The departure. When the bride and groom leave the reception for their honeymoon, the guests usually gather outside to throw rice or confetti and wave goodbye. Get there early so you have a good shot of the wedding couple entering the car and driving away.

That's it! You've shot the perfect wedding video, and whatever isn't perfect can be edited away. Make some nice titles, and you have given the wedding couple a wonderful gift to treasure. Now you know why wedding videographers get so much money for this.

Creating an Online Seminar Web Course

The web course example that follows was created to be used with or without supporting online content and other resources. The online class or course using the series of video clips is formed around a real-time or synchronous online meeting. The script of the course was prerecorded using a consumer digital video camcorder, edited and exported as RealVideo files. These files were then imbedded into the web pages of the course. The website also featured the use of Microsoft NetMeeting to create real-time class meetings to discuss the video content and other topics. The meeting may be text-only chat or audio/video conferencing. Using simple consumer video equipment, anyone mastering the earlier chapters of this book is capable of this kind of project.

A Video Web Course: Fundamentals of Historic Building Preservation

This web course is a series of 12 Internet distance-learning study units, consisting primarily of online video imbedded into web pages, covering the fundamentals of historic building preservation. They have a brief text and photographic introduction to the subject being covered and have eight to twelve minutes of video presentation by historic preservation architect Walt Marder, with the Florida Division of Historical Preservation. Walt gives a brief overview of the topic. In addition, there is a short multiple-choice quiz to help reinforce the key items covered in the unit. A list of available study or reference resources is included in each unit along with links, placed on each web page where the video is imbedded, to appropriate preservation briefs and other websites. The current units will be expanded over time and additional units at the fundamental level and more advanced ones will be added as time and funding permit. The video study units are:

- Introduction
- Identifying Characteristics
- Definitions
- Wood
- Metals
- Masonry
- Codes
- Anachronisms

- Archeology
- New Additions
- Maintenance and Cleaning

These study units, combined with an instructional program, can form the basis for an Internet or CD-ROM–based distance-learning course on the fundamentals of historic building preservation. The course, with additional resources added, can become more advanced in both theory and practice and can be used at the college level. The units will be utilized by the Florida Division of Parks and Recreation to introduce the park ranger staff to the basics of preservation maintenance.

A brief look at the major elements of the units along with a screen capture of each element follows, starting with Figure 19-1.

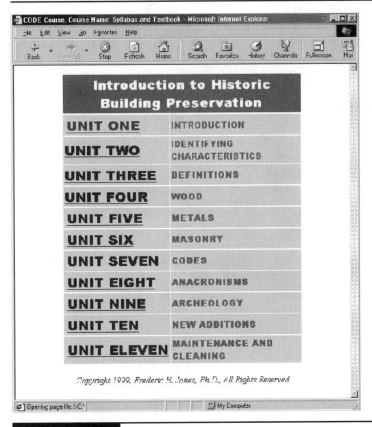

FIGURE 19-1   Main course menu

The main course menu, shown in Figure 19-1, links to the 11 current historic building preservation study units. Additional study units and other resources are planned for the future. These include a preservation bibliography, glossary, and a database of historic structures and architectural elements.

Each study unit has one or more pages of introductory and supporting materials (see Figure 19-2). This includes guidelines, references, links, bibliographies, and other textural and graphic information.

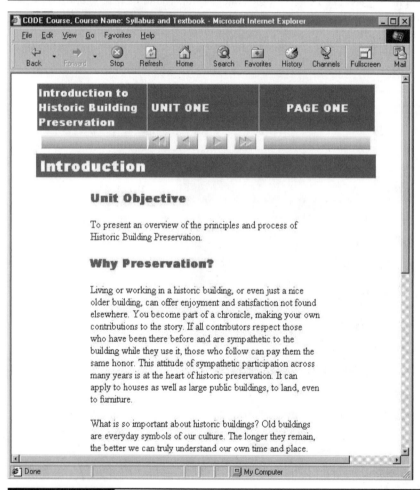

FIGURE 19-2   Unit One page

Each of the current study units has a brief video overview by preservation architect Walt Marder of the Florida Division of Historical Resources (see Figure 19-3).

Each study unit contains a brief quiz to help reinforce the key concepts covered in the unit, as shown in Figure 19-4. These may be multiple-choice or true/false questions and are graded instantly by the computer. These quizzes can be expanded by an instructor for use in distance learning course offerings.

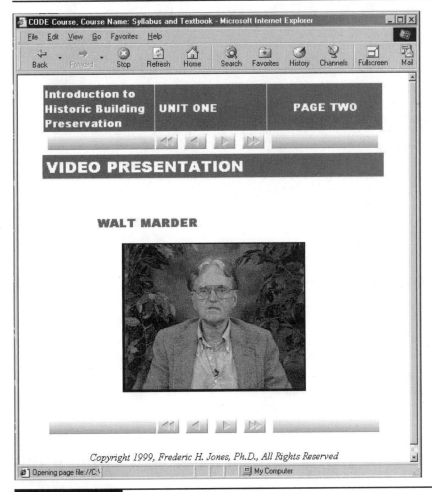

FIGURE 19-3   Unit One video page

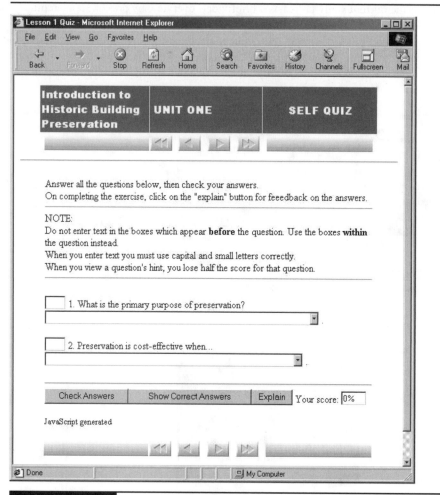

FIGURE 19-4   Unit One quiz

Other Ideas for Using Your Talents as a Videographer

I could go on forever with ideas about ways to use your video camera and your talents, but you get the idea. Following is a small list of other projects you might consider:

Videodiary Have a new baby? Make a video of yourself and your spouse for the baby, talking about your feelings, your experiences, your new insights, your hopes and dreams for your new child. This makes a nice gift for a grown child who can look back and see and hear their parents at the very beginning of the family.

School projects   Help your kids learn how to use the video camera and make video projects for school. My daughter made a small "family story" video about her family's Thanksgiving traditions for a class on family cultural heritage.

Mini-movie or documentary   You've got the skills and the talent! Go for it!

Summary

The possibilities are endless. By putting together all the things you have learned—shooting interesting and well-coordinated shots, good sound, music, voiceover, effects and titles, editing with an eye to detail—you can make video projects that will surprise even you. This is an enormously creative and popular media form, but one that too many people fear to tackle because it sounds hard. Now you know the truth! Feel free to add to the list!

Part V Appendix

Appendix

Use Windows XP Movie Maker and Windows Media Video 8

Microsoft Windows XP offers a much more complete set of digital video and audio tools and capabilities included with the system itself than in earlier versions of Windows. You can now import and export digital video and do simple edits resulting in a movie with the built-in Movie Maker software. It is very simple, but that is a virtue for the novice. It is extremely easy to use and the results are excellent. It can be used to create Web video and video attachments for email as well as simple video projects. The new Windows Media Player has more features than its predecessors do. You will, of course, still need a FireWire digital connection to use a digital video camcorder with your computer (covered in Chapter 7). Windows XP simplifies some of the video importing and exporting functions in the software utilities covered in Chapters 8 and 14, and the built-in utilities may be used as a substitute for third-party software packages to transfer video clips to your computer. You will need to check with your camcorder manufacturer for support information.

This appendix assumes that you have certain skills or have read earlier chapters of the book. The skills and the chapters that cover them are as follows:

- *Using your digital camcorder (Chapter 1)*
- *Shooting quality video (Chapters 2 and 3)*
- *Video lighting (Chapter 4)*
- *Video sound (Chapter 5)*
- *Production preplanning and scripting (Chapter 6)*
- *Selecting and installing an IEEE 1394 or FireWire connector in your computer (Chapter 7)*

Windows Media Audio and Video 8

With the release of Microsoft's Windows XP operating system, which includes more extensive digital video and audio support, the latest version of their audio compression scheme, the Windows Media Audio 8 format, is also available. It provides essentially the same audio quality as typical MP3, but with a file that is about one-third the size. This results in three times the amount of music for the same storage and much faster download times. When streaming, the Media Audio 8, which can play audio at multiple speeds, delivers audio at 64 Kbps, comparable to CD quality, and at 96 Kbps and above, which equals original audio sources.

The Windows Media Video 8 format is designed for both streaming and downloads, and provides compression efficiency that results in near-VHS quality (320 x 240 pixel resolution at 24 frames per second) at rates as low as 250 Kbps. This means that your streaming video is nearly as good as broadcast TV at the low end of DSL/cable connection speeds. Media Video 8 can also deliver near-DVD quality video at rates as low as 500 Kbps. (DVD quality is 640 x 480 pixel resolution at 24 frames per second.) This means that near-DVD quality is delivered at typical DSL/cable connection speeds.

Windows Media Encoding Tools

In addition to the new video and media tools for novices, Microsoft has also provided a number of advanced tools to encode your video in the new Windows Media 8 format. These are included in the release version of Windows XP and are available from software developers such as Adobe and Dazzle that support the format from their video editing software packages. The basic encoder is the Windows Media Encoder 7.1 tool, which supports live capture for Windows Media Video 8 streaming, archived encoding for video streaming, and live or archived audio encoding for streaming and downloads. Media encoding tools are a bit beyond the scope of this book, as these tools are command-line driven and are not meant for the novice user. If you are interested in pursuing this further, you can download both the tools and the documentation on their use from the Microsoft website.

There are other professional tools that are available and can be reviewed at the Microsoft Windows Media website, http://www.microsoft.com/windows/ windowsmedia/. For Windows Media Audio and Video 8 in particular, go to http://www.microsoft.com/windows/windowsmedia/WM8/.

Windows Movie Maker for Windows XP

The Windows Movie Maker component of Windows XP is a basic video and multimedia editor that can capture, edit, and record video and audio files directly from and to DV camcorders, CD-ROM, email, and the Internet. Windows Movie Maker works with IEEE 1394 FireWire ports and most DV camcorders to import directly into the computer. It automatically detects a camera and provides the option of immediately capturing footage into Windows Movie Maker in real time. It supports both analog and digital input and works with digital video cameras, analog video recorders, and VCRs if your computer has the appropriate digital or analog input devices (as discussed in Chapter 7).

There are several neat features of Movie Maker, including:

Automatic shot detection   Movie Maker automatically detects when scenes change and provides a visual display of thumbnails from every scene. This allows you to find the scenes you want without fast-forwarding through the entire video.

Drag and drop scenes   You can edit a movie by dragging thumbnails or small icons that represent each scene from the workspace onto the timeline.

Trim to the highlights   Clips can be previewed and then trimmed or divided into segments on the workspace or the timeline.

Easy sound assembly   You can record a voiceover or music and then drag it onto the timeline.

Improved video compression   Movie Maker imports footage in Windows Media Video format. This format allows for storing 100 hours of high-quality video (256 Kbps) in

11.5 GB (about nine hours per gigabyte) of hard drive space. A three-minute movie at 100 Kbps is just over 2 MB.

Sorting by subject   Movie Maker's shot-detection feature allows you to easily sort a video clip library by subject.

Windows Media Player   Movie Maker videos can be played in any media player that supports Windows Media format, including Windows Media Player.

Using Windows XP Movie Maker

Windows Movie Maker requires a hardware connection between your camcorder and your computer. If you are using a digital video camcorder, you will need an IEEE 1394 connection. If you are using analog video, you will need some kind of analog video capture board or port (both of these options are covered in Chapter 7). If the hardware connections are provided and properly installed, you can proceed directly to Movie Maker to import and edit your video and audio files. The main screen of Movie Maker is shown in Figure A-1.

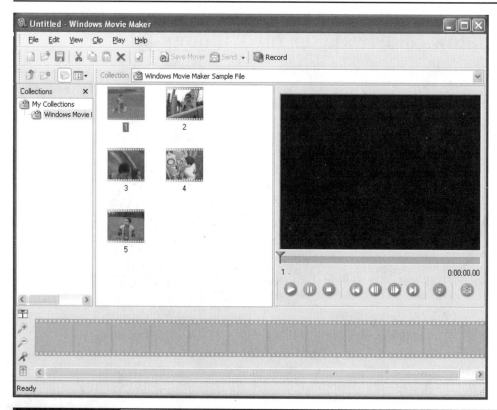

FIGURE A-1   Movie Maker main screen

App

> **NOTE** *You may need to install new or additional drivers to use your capture boards with the new Windows XP operating system. A driver is a small software utility that allows your video equipment to work with the capture card in your computer. Check the website of the manufacturer of your video equipment to make sure the appropriate driver is available. If it isn't, you may need to call or write the manufacturer to find out when it will be available. Windows XP comes with many IEEE 1394 interface card drivers already installed.*

Transferring from Camera to Hard Drive

Window's Movie Maker is simple to use and understand. The first thing you need to do is transfer your video footage to your hard disk so you can begin working with it.

1. Make sure your video devices are properly attached to your computer. After your video devices have been configured, open Windows Movie Maker and click Record on the toolbar. You will initially be presented with the Record dialog box (see Figure A-2) and a number of choices, including:

   ■ Which device you want to record from, if more than one is connected

   ■ Record audio only, video only, or both

   ■ A recording time limit

   ■ A general quality level

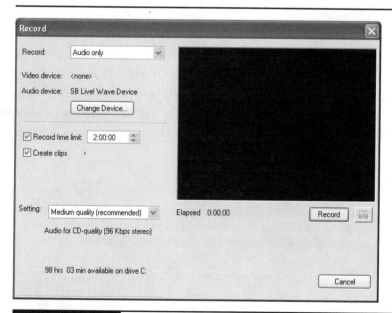

FIGURE A-2   Movie Maker Record dialog box

2. Movie Maker has a feature that can save time when importing your video clips: the Create Clips shot detection function. This automatically creates separate clips when a different frame type is detected. This occurs when you shoot a different subject with your camcorder or move the camera to a different shot. To enable this feature, select Create Clips before you begin recording.

TIP   *The Create Clips function is a real help to beginning video makers. It eliminates the need to fast-forward and fast-reverse using the camera controls, attempting to find the beginning and end of clips manually. A word of caution, however: the software doesn't know what you consider to be a clip. It just extrapolates from the change in the background of the shot you are making. If the shots are similar, it won't save them all consistently as clips. You will still need to edit the automatically selected clips into smaller segments. You can also wind up with too many clips if you intend to include longer segments just as recorded. Experiment with the feature to determine if and when it is useful for your project.*

3. When you have selected your settings, click Record. The word "Recording" will blink, indicating that you are recording. The elapsed time of your current recording will show next to it. When your recording has begun, the Record button becomes a Stop button. Click Stop when you want to stop recording.

NOTE   *You can import files other than digital or analog video into your Movie Maker timeline including audio, still photographs, PowerPoint files, and others. Here is a list of file types:*

- *Video files with an .asf, .avi, or .wmv file extension*
- *Movie files with an .mpeg, .mpg, .m1v, .mp2, .mpa, or .mpe file extension*
- *Audio files with a .wav, .snd, .au, .aif, .aifc, .aiff, .wma, or .mp3 file extension*
- *Windows Media-based files with an .asf, .wm, .wma, or .wmv file extension*
- *Still images with a .bmp, .jpg, .jpeg, .jpe, .jfif, .gif, or .dib file extension*
- *Microsoft PowerPoint files and individual slides with a .ppt extension*

4. To add one of the file types to your movie choose File | Import (see Figure A-3). Find the file you want to add, and click Open.

The edit timeline in Windows Movie Maker runs along the bottom of your screen, below the collections area and the monitor. There are two working views for editing: the storyboard view and the timeline view (see Figure A-4). You can use either or both in editing. The storyboard view allows you to create a sequence of your clips. The timeline view lets you control how the clips interact both with the audio track and with each other.

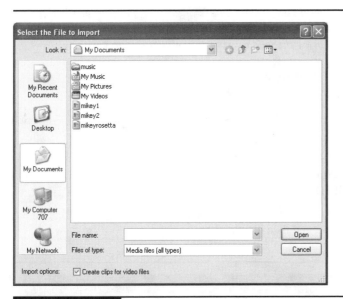

FIGURE A-3   Movie Maker Import dialog box

FIGURE A-4   Movie Maker view selection

TIP

Use the storyboard view when you are organizing your project initially, then switch to the timeline view for final addition of transitions and editing adjustments. This will speed up your overall editing process.

Using the Storyboard View

1. To use the storyboard view (shown in Figure A-5), drag clips from the collections area onto the storyboard in the order you would like them to appear in your movie. You can rearrange the clips by dragging them to a different location on the storyboard.

2. You can remove a clip from the storyboard by clicking on the clip on the storyboard and then selecting Edit | Delete.

3. You can preview your assembled video in the project monitor by clicking on an empty area in the workspace, and then clicking Play.

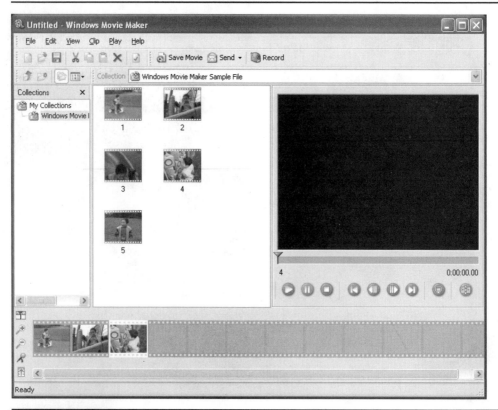

FIGURE A-5   Movie Maker storyboard view

When you are doing more detailed editing such as trimming the ends of the video clips or adding a better transition or an audio track, you will be better served by using the timeline view.

Using the Timeline View

The timeline view expands the view of each clip with a series of numbers superimposed on the clips that indicate the duration of each clip. The timeline view also shows the interaction between clips, and between clips and the audio track. The audio track is indicated by the bottom bar in the workspace (see Figure A-6). Zoom In and Zoom Out buttons are provided to the left of the workspace to allow you a more detailed view of your clips.

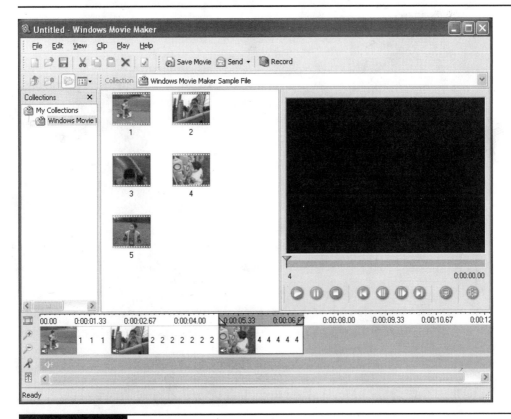

FIGURE A-6   Movie Maker timeline view

When you are using the timeline view, transitions can be added in addition to cuts that result from simply placing one clip by the next. You might add a cross-fade or lap dissolve in which one scene fades out while the next scene appears behind. This can be created by dragging the second of two adjacent clips to the left so that its icon overlaps the first clip. The shaded area indicates the length of the transition.

A fade-to-black dissolve can be achieved by creating a black still image in your paint software package and importing it as a clip to your timeline, then adding a cross-fade transition from a video clip to the black image.

You can add a voiceover narration track by recording an audio track using the Record Narration button located on the timeline (see Figure A-7). This feature lets you add an audio track that will be keyed exactly to the clip sequence. You can watch the video clip while you add narration to it.

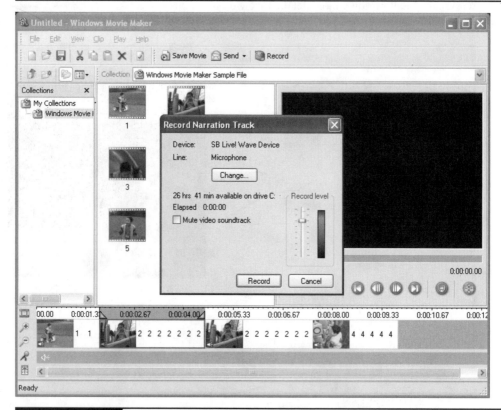

FIGURE A-7
Movie Maker Record Narration Track dialog box

Make sure that you plan your voiceover or narration ahead of time by previewing the clips and making notes. This will speed the overall process and ensure a better quality sequence. The Record Narration feature makes doing and redoing of narration easy and less frustrating than more complex editing software systems.

> **NOTE**
>
> *Audio clips can be added to your timeline using the same process that is used to add video clips. The cross-fade effect does not work on audio tracks. Overlapping audio tracks results in mixing the two tracks and allowing them to play simultaneously.*

Clicking the Audio Levels button next to the timeline allows you to control the audio level of a clip (see Figure A-8). You can increase the audio level by dragging the slider bar to the right.

FIGURE A-8   Movie Maker Audio Levels control

Saving Your Movie

To save your finished movie, click File | Save from the top menu and give your file a name (Figure A-9). You can choose various compression and file types when you save. These choices can be very complex if you are saving to formats other than DV. For more information refer to Chapter 8.

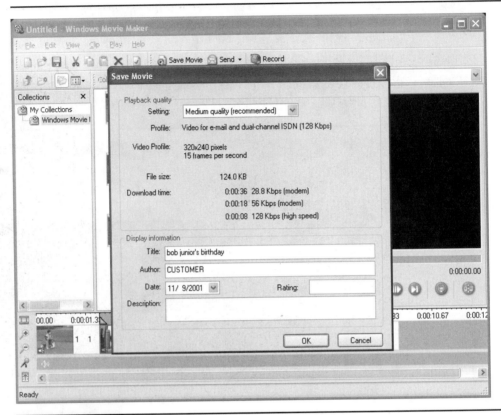

Movie Maker Save Movie dialog box

View Your Movie Using Windows XP Media Player

When your movie is complete, you can view it using the Windows Media Player (see Figure A-10). It is a much more sophisticated software tool than prior versions in Windows and allows more control over your audio and video collection both on your hard drive and on the Internet.

FIGURE A-10   Windows XP Media Player

Summary

Have fun with the Windows XP media features. They can come in handy when you need to create a quick clip of video for the Internet or to email a friend. They will not, however, replace more serious video and audio editing software for larger projects.

Index

INTERNATIONAL CONTACT INFORMATION

AUSTRALIA
McGraw-Hill Book Company Australia Pty. Ltd.
TEL +61-2-9417-9899
FAX +61-2-9417-5687
http://www.mcgraw-hill.com.au
books-it_sydney@mcgraw-hill.com

CANADA
McGraw-Hill Ryerson Ltd.
TEL +905-430-5000
FAX +905-430-5020
http://www.mcgrawhill.ca

GREECE, MIDDLE EAST, NORTHERN AFRICA
McGraw-Hill Hellas
TEL +30-1-656-0990-3-4
FAX +30-1-654-5525

MEXICO (Also serving Latin America)
McGraw-Hill Interamericana Editores S.A. de C.V.
TEL +525-117-1583
FAX +525-117-1589
http://www.mcgraw-hill.com.mx
fernando_castellanos@mcgraw-hill.com

SINGAPORE (Serving Asia)
McGraw-Hill Book Company
TEL +65-863-1580
FAX +65-862-3354
http://www.mcgraw-hill.com.sg
mghasia@mcgraw-hill.com

SOUTH AFRICA
McGraw-Hill South Africa
TEL +27-11-622-7512
FAX +27-11-622-9045
robyn_swanepoel@mcgraw-hill.com

UNITED KINGDOM & EUROPE (Excluding Southern Europe)
McGraw-Hill Publishing Company
TEL +44-1-628-502500
FAX +44-1-628-770224
http://www.mcgraw-hill.co.uk
computing_neurope@mcgraw-hill.com

ALL OTHER INQUIRIES Contact:
Osborne/McGraw-Hill
TEL +1-510-549-6600
FAX +1-510-883-7600
http://www.osborne.com
omg_international@mcgraw-hill.com